# LEP – The Lord of the Collider Rings at CERN
## 1980–2000

Herwig Schopper

# LEP – The Lord of the Collider Rings at CERN 1980–2000

## The Making, Operation and Legacy of the World's Largest Scientific Instrument

With a Foreword bei Rolf-Dieter Heuer

 Springer

Prof. Dr. Herwig Schopper
CERN
1211 Genève 23
Switzerland
herwig.schopper@cern.ch

ISBN 978-3-540-89300-4        e-ISBN 978-3-540-89301-1
DOI 10.1007/978-3-540-89301-1
Springer Dordrecht Heidelberg London New York

Library of Congress Control Number: 2009920206

*Cover design*: WMXDesign GmbH, Heidelberg

Printed on acid-free paper

Springer is part of Springer Science+Business Media (www.springer.com)

# Foreword

Which better year than 2009 to publish a book describing the history of LEP? At the time of writing these lines the LEP experiments are going to publish their last scientific papers eight years after the end of data taking and more than thirty years after the first studies for a large electron-positron collider which eventually became LEP. It is now the right time to look back how everything has started and how this scientific project made its way so successfully from its birth to its last, scientifically dramatic years. And who else would be better placed to write the story, and not only the history, of LEP than Professor Herwig Schopper, Director General of CERN from 1981 to 1988, who not only obtained the approval of LEP but also brought its construction to a successful end during his term of office. With all his insights into the scientific, technical, political, financial and managerial aspects and challenges of such an endeavour he does not forget the human aspects; and the personal recollections and anecdotes expressed by the author help to understand the reader what it means to carry such a project through. Without LEP, we would not have the same knowledge of particle physics to day: LEP has had a strong impact on the experimental 'discovery' of the Standard Model. But LEP has not only pushed the frontier of knowledge, it also advanced the frontier of technology, of computing and of worldwide collaboration. This book is also a reminder to everybody about the duration of our projects, about the stability needed to carry them through, but also about the excitement they give to us! The exciting times of LEP which shaped generations of physicists, are over – but more are ahead of us with the turn-on of the Large Hadron Collider (LHC). LEP has been making way for this new global project of particle physics at CERN, which will push further ahead our knowledge of the microcosm and of the early universe. But LEP will live on in physics text books and as a model for international collaboration beyond all political boundaries. As stated by the author, with LEP and its international exploitation, CERN has become a laboratory serving the worldwide community of elementary particle physicists. This book addresses the specialists as well as the interested public, you don't have to know particle physics but you will learn much. Enjoy reading as much as I did.

CERN, February 2009
*Rolf-Dieter Heuer*
Director-General CERN

# Prologue

Sometimes the Large Electron–Positron Collider (LEP) and its successor in the same tunnel, the Large Hadron Collider (LHC), which is scheduled to become fully operational in 2009, are compared to cathedrals built in the Middle Ages. They have some aspects in common, such as belonging to the largest objects created by man, requiring a considerable collective effort, applying the most advanced techniques available at their time and devouring a large fraction of the GNP. They expressed in many respects the spirit of their age, although certainly some people considered them as useless from a practical point of view. Hence, it may be of interest to permit a glance behind the scenes of the making of LEP, the largest scientific instrument built by man.

This is not a report written by a historian, but rather by somebody who was directly involved in the realization of this spectacular project at a crucial position. I am also sufficiently old that I do not have to care too much about personal sensitivities, but rather can tell the 'truth'! Historians base their findings mainly on documents or interviews. My experience is that this does not give a full history of the motivations, difficulties and personal involvements. Indeed, the real decisions are mostly not taken during official meetings, but in coffee breaks, at dinners, in the corridors or during telephone calls. When a group of historians was charged with writing the story of the foundation of CERN, their leader came to see me one day and announced that they wanted to base their report only on documents, since in some interviews with the founding fathers of CERN they recognized that the old gentlemen had fading memories partly contradicting each other. My advice was that they should immediately return to listen to the oral tales since in my mind the contradictions had nothing to do with forgetfulness but rather with different points of view. The accounts of participants at a meeting from the previous day would also be contradictory. On this occasion I realized how difficult it is to write an unbiased historical report.

About 20 years have passed since the building of LEP. I did not write the history of LEP earlier since I have been very busy with other activities[1] and hence my

---

[1] President of the German and European Physical Society, consultant to UNESCO, President of SESAME Council and others.

time was limited in spite of my being formerly retired. Because of the time that has elapsed, some recollections might have become somewhat distorted in my memory and I apologise for any omissions or mistakes which might have crept into my story. On the other hand, some detachment seemed useful to have a less biased view, and some information that was confidential at the time of the construction of LEP can now be made public. In addition, the final analysis of LEP experiments came to an end only recently, thus permitting a full evaluation of the impact of LEP on particle physics. Finally what we called phase LEP 3 (now called the LHC) started operation in 2008. When the parameters of the LEP tunnel were chosen, the choice was made with the view of a hadron collider being in the same tunnel. This vision is becoming a reality now and thus LEP is part of the early history of the LHC. Sometimes it was also amusing to see that some of the problems which we experienced with LEP appeared again with the LHC, but had been forgotten in the meantime.

A project such as LEP has many different aspects, technical, scientific, managerial and political, and human problems are always a major part of any history. I have tried to cover all those aspects to give a complete picture. Therefore, most readers will not read this book from the first page to the last, but they will rather select those chapters which preferentially raise their interest. But I hope they might come back and try those parts which are somewhat more remote from their usual background.

Why was this book written at all? This less formal but hopefully informative source of the history of LEP will provide some information which otherwise might be lost, although this account is not intended to replace a professional report by historians. More importantly LEP (and including its last phase, the LHC), the largest facility for basic research ever built by humankind, has become a kind of symbol for the efforts humanity is prepared to make to explore fundamental questions related to understanding for the basis of human existence and our view of the world. One of the frontiers of our knowledge concerns the infinitely small and with LEP we could penetrate deeper into the structure of the microcosm. The other rim of knowledge concerns the opposite, the infinitely large, the universe. Recently the study of the infinitely small and that of the infinitely large have become intimately intertwined – a most fascinating development. Of course, a detailed description of this topic is not the objective of this book,[2] although some indications will be given.

LEP has proven that Europe can become a worldwide leader in science and technology when efforts are combined. Finally, the realization and exploitation of LEP involved scientists and engineers, administrators and politicians from countries with different traditions, mentalities, religions and political systems. Cooperating closely together has proven that science can be a tool for creating better understanding and building of mutual trust in the spirit of the UNESCO slogan 'Science for Peace', one of the main objectives implemented at CERN right from its foundation.

---

[2] Sometimes a third frontier is supposed to be complexity, which might be connected with biological life.

# Contents

# Chapter 1
# Why LEP and Why at CERN?

## 1.1 Is Curiosity-Driven Research Justified?

LEP and its successor in the same tunnel, the LHC, are the largest research facilities ever built, but they provide no immediate visible benefit for the economy. Indeed a machine such as LEP serves to explore the mysteries of the microcosm, a domain which is accessible in detail to only a handful of experts. Does this justify the expense and human efforts to realize such a facility? Similar questions may also be raised with respect to large telescopes or satellites exploring the deep universe. Two main answers can be given.

First, during the past 200 or 300 years, curiosity-driven research has completely changed our concept of the world (our *Weltbild* in German) and the position of man in nature. If we are not scared anymore by thunder as an expression of an angry god, it is due to basic research explaining it as a natural phenomenon. Our whole perception of our environment as a rational and unmystified world has been established by the results of fundamental research, which at the beginning is always accessible only to experts, but which is eventually taught in schools. We know now that the universe and with it the earth are much older than a few thousand years, we know that electricity and magnetism are two sides of one basic force and we have learned how life has developed through evolution. Such and similar information has no direct practical application, but it is essential for the understanding of the position of humankind in the universe.

This, of course, also raises questions about the relations between science and religion. This important topic cannot be considered in more detail here, but at least in Chap. 11 I shall report that Pope John Paul II agreed with me that science and religion are two ways of grasping reality, being complementary and not contradictory.

The second justification lies in the fact that from curiosity-driven research the most innovative applications resulted. All progress of modern society is based on such inventions and only the application of the results of basic research has made it possible to suppress slavery and bring welfare and culture to a large part of mankind. This process will be discussed more explicitly in Chap. 9.

To penetrate into the microcosm and get new knowledge about the fundamental constituents of matter and the forces which act between them, one needs instruments

H. Schopper, *LEP – The Lord of the Collider Rings at CERN 1980–2000*,
DOI 10.1007/978-3-540-89301-1_1, © Springer-Verlag Berlin Heidelberg 2009

which extend our senses to extremely small dimensions, such as 'supermicroscopes'. The smaller the objects are which we want to explore, the higher the energies must be for us to 'see' them.[1] One might also put it another way. If we want to find the smallest building blocks of matter, we have to break them up into smaller and smaller pieces: molecules into atoms, atoms into nuclei and electrons, atomic nuclei into protons and neutrons (which together with similar particles are called 'hadrons') and finally hadrons into quarks (Fig. 1.1). But it turns out that the smaller the building blocks are, the harder they are to detect and hence higher energies are necessary to break them up. Hence, most of our knowledge about the microcosm has come from the ability to produce high-energy collisions between particles by means of accelerators or storage rings and to observe the subsequent behaviour by means of sophisticated detectors.

It is a tradition in elementary particle physics that the big laboratories look well ahead in their planning, involving potential future users from universities and national laboratories. This has sometimes provoked criticism from the outside that elementary particle physicists are insatiable in jumping too fast from one project to

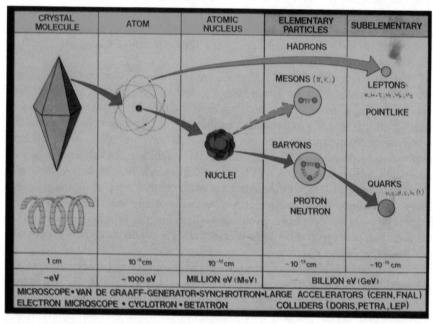

**Fig. 1.1** Constituents of matter. The discovery of ever-smaller building blocks of matter is indicated. Their sizes and the energies (given in units of electronvolts) needed to break them up are shown, as well as the instruments (microscopes, accelerators, colliders) used for their discovery

---

[1] As is well known from the optical microscope, the wavelength of light must be shorter than the dimension of the objects to be investigated. According to quantum mechanics, a wavelength can also be attributed to particles. The higher their momentum, the shorter the wavelength.

the next. However, taking into account that the realization of a large project from first ideas to starting operation takes more than 10 years, such a policy has the advantage that new projects are well prepared, both technically and scientifically, with the consequence that they can be realized within the planned timescales and budgets and that they are fully exploited.[2] LEP is a particularly good example for such a procedure.

## 1.2 Colliders Surpass Accelerators

In accelerators, sophisticated combinations of electric and magnetic fields can be used to confer high energies to charged particles, usually protons (the nucleus of the hydrogen atom) or electrons, which are both stable and easy to liberate from atoms. The highest energies can be achieved with so-called synchrotrons, which are accelerators where particles orbit in a ring of magnets many times per second and high-frequency electric fields in synchrony with the orbiting particle (hence the name) 'push' them repeatedly to increase their energy. When the desired energy has been reached, the beams can be ejected from the magnetic ring and aimed at 'fixed targets' outside the accelerator.[3]

The advantage of extracted beams sent to external targets is that high beam intensities can be achieved. In addition, elusive or short-lived particles (e.g. mesons, muons, neutrinos) may be produced in the external target and secondary beams of such particles can be derived for further use. Extracted beams and secondary beams offer many combinations of incident and target particles. This permits the investigation of building blocks of matter as if they were illuminated with light of different colours. However, there is no gain without loss! The main disadvantages stem from the fact that the target particles cannot be fixed at their positions and, in particular, at very high energies the particle that is hit yields to the impact of the incident particle and picks up momentum, very much like a billiard ball hit by another billiard ball. Thus, a large fraction of the kinetic energy of the incident particle is converted to recoil energy and is lost because of the collision itself.

To avoid these recoil losses it is much more efficient to shoot two particle beams head-on against each other. It is well known from daily life that much more damage is produced when two cars collide frontally instead of hitting a fixed obstacle. This difference is even greater when the rules of the 'special theory of relativity' (speed

---

[2] As for every rule there are also exceptions to this one. The most spectacular failure was the large hadron collider Superconducting Super Collider (SSC) in the USA, the building of which was stopped by the US Congress after several billion US dollars had been spent. The reasons for this failure are complex [1]; the LHC also had some financial difficulties, mainly owing to the enormous challenge posed by the superconducting magnets cooled with suprafluid helium, a completely new technology for such a large project.

[3] The development and functioning of accelerators and colliders are described in a more popular way in [2, 3].

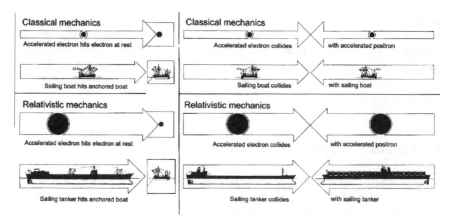

**Fig. 1.2** The difference between classical and relativistic mechanics in collisions. In classical mechanics the effects in head-on collisions are worse than those for collision with an object at rest, but in relativistic mechanics head-on collisions are disastrous. The size of the points indicates the mass of the particles

of particles close to the speed of light, 'relativistic mechanics') have to be applied, according to which the mass of an object increases drastically when its speed approaches the speed of light. This is illustrated in Fig. 1.2.

These facts led to the invention of particle 'colliders', facilities in which particles collide head-on. These colliders have made accessible effective collision energies which could never have been produced by 'normal' accelerators and they have thus opened new windows for the exploration of the microcosm.

Magnetic rings as used in synchrotrons can be employed to first accelerate particles to high energies by applying an electric field and then switching it off when the desired energy has been reached.[4] Subsequently the particles are stored, i.e. one lets them run around in the magnetic ring for hours. One sometimes speaks, therefore, of 'storage rings'. If particles of the same kind (i.e. carrying the same electric charge) are to be brought into collision, two rings (with opposite magnetic fields for the particles circulating in opposite directions) which intersect in a number of places must be used. Different geometries are possible, but the rings can intersect only at a certain angle and hence the collisions will not be precisely head-on. Only one such facility was realized in the past, the Intersecting Storage Rings (ISR) at CERN colliding protons with energies of $2 \times 30\,$GeV. The largest proton–proton collider ever built was constructed at CERN in the LEP tunnel, the LHC, with an energy of $2 \times 7,000\,$GeV and which started operation in 2008 (see Chap. 14).[5]

---

[4] As will be explained later in the case of electrons, the accelerating field cannot be switched off completely but has to be maintained partially to compensate for energy losses.

[5] A proton–proton collider for $2 \times 400\,$GeV called 'ISABELLE' was started at the Brookhaven National Laboratory in 1979 but never came into operation.

An alternative possibility is to produce collisions between particles and their antiparticles,[6] in which case one magnetic ring suffices since particles with opposite electric charges can circulate in the same magnetic field in opposite directions and they will automatically collide head-on. To avoid collisions taking place all around the circumference, the ring is not filled uniformly with the circulating particles, but they are concentrated in packets (by experts called 'bunches') and therefore collide only in certain places where the packets happen to meet. By controlling the timing of these 'bunches', one can make sure that they collide only at places where the experiments are located.

So far matter–antimatter colliders have been realized in two versions, electron–positron and proton–antiproton colliders. The first electron–positron colliders were built in Italy at the Frascati Laboratory near Rome and many more were realized in Europe, the Soviet Union, the USA, Japan and China since the production of positrons, the antiparticles of electrons, is easy thanks to their small mass. Low energies[7] (minimum 1 MeV) are sufficient to produce them in large quantities; however, antiprotons are much more difficult to produce in sufficient quantities because of their high mass, and energies of several gigaelectronvolts are required for their production. Suitable machines were realized in the past only at CERN in Europe and at Fermi National Accelerator Laboratory (FNAL) in the USA.

Since electrons and their antiparticles are, as far as we know, fundamental (which means that no internal structure has been discovered), their collisions produce very clean events which are easy to analyse. On the other hand, protons and their antiparticles have a complicated structure (consisting of quarks and gluons; see Chap. 8) and hence their collisions result in complicated events. The basic collision is between two quarks or a quark and a gluon and many other particles not directly involved in the collision confuse the interpretation of the event as 'observers'. The 'real' collision energy is not that corresponding to the sum of the energies of the two colliding protons since the energy of a proton is shared between its different constituents. On average the effective energy for a collision between proton constituents is about one third of the total energy of the protons. Protons, however, are easier to accelerate to high energies, whereas electrons emit synchrotron radiation and therefore the achievable energies are much lower. Both kinds of machines have, therefore, their advantages and drawbacks and they complement each other. The technical challenges that need to be overcome to reach high energies are quite different for electron and proton machines. The maximum energy for electrons is limited by the emission of so-called synchrotron radiation (see Chap. 6) and this loss has to be compensated for continuously by very powerful high-frequency acceleration systems. For protons, on the other hand, which emit very little synchrotron

---

[6] Each particle in nature has its antiparticle, which is a kind of mirror image having opposite electric charge and other properties but the same mass (see Chap. 8).

[7] When dealing with elementary particles, it is useful to use a special unit for energy, the electron-volt (eV), or better $1 \times 10^6$ eV = 1 MeV, or 1 GeV = 1,000 MeV. Even 1 GeV is an extremely small amount of energy, just sufficient to lift an ant by 1 mm. In particle physics it is the concentration of energy in a small volume which matters.

radiation, powerful magnets must be applied to keep them on circular tracks. In both cases superconductor technologies have to be applied, as will be discussed in Chap. 6.[8]

All colliders have the advantage that particles may be stored for many hours in magnetic rings, with the particles circulating in opposite directions and interacting at a few points where the experiments are located. If the particles do not collide at their first encounter, they return for another encounter after a short time and a collision may then occur. With the two beams passing through each other many million times per second, the probability for collisions is greatly increased, thus providing a high number of collisions per second. This number depends on the technical parameters of the facility (see Chap. 6) and to maximize it is part of the art of designing the machine.

At CERN, the large electron–positron collider LEP, which had a maximum beam energy of 110 GeV, was built in the 1980s, but it has been replaced by the proton collider LHC, which has a maximum beam energy of 7,000 GeV.

Large detection systems surround the locations where the particles collide ('interaction points'). The detectors attempt to observe all the products of the collisions and to obtain information on their directions, their energies and the type of emerging particles. This requires sophisticated detectors with extremely high accuracies to observe particles in space and in time. (Considerable technological developments are necessary for this purpose and the technologies developed for particle physics have found many applications in other fields, in particular in medicine.) Having observed the various particles produced in the collision, we can then try to find out what really happened. Such an interpretation is similar to detective work where one has to conclude from the fingerprints who the culprit was. To perform such analysis requires considerable computing power for the data acquisition and the analysis. Since most such detection systems are developed and built in international cooperation with scientists and engineers distributed all over the world, it was also necessary to develop new kinds of data networks to make the registered data available to all users. Hence, it is not surprising that the World Wide Web was developed at CERN for the benefit of the LEP users by Tim Berners-Lee and Robert Cailliau. It has now become a ubiquitous element of communication and has changed our society. Unfortunately, nobody foresaw the importance of this invention in the 1980s and we did not take out a patent (see Chap. 9 for more details).

## 1.3 The Stage for a World Facility

After World War II, there was competition between the USA and Europe to construct facilities for elementary particle physics, and a number of accelerator and collider ring types were duplicated on both sides of the Atlantic. In parallel to this

---

[8] For linear colliders see Chap. 2.

transatlantic competition, independent objectives were pursued by various European nations and US laboratories. The fragmented Europe recognized early that with the growing cost of the facilities international cooperation was needed. The first consequence was the foundation of CERN in 1954 [4] and the setting up of the European Committee for Future Accelerators (ECFA), which tried to coordinate the efforts in Europe.

In the 1970s it became obvious that further penetration into the microcosm required facilities which seemed to be realizable only on a world scale. To discuss this issue, a topical seminar was held on 3–7 March 1975 in New Orleans, USA, with leading scientists from all the countries involved in particle physics. At that time I was director of DESY, the particle physics laboratory in Hamburg, Germany, and in that function I had also been invited. A number of topics were discussed, such as future national projects and better ways of international cooperation, but the main item was the question of how to build a large facility beyond the capabilities of individual regions. The main outcome of the meeting was an agreement that the next big facility, a Very Big Accelerator, the 'VBA', should be considered and if possible realized as a world machine and it was suggested that an international study group should be established with its headquarters at CERN [5]. Although no definite proposal for the VBA was made, the possibility of a large electron–positron collider received priority. The outcome of this meeting was discussed at CERN in meetings[9] of the Scientific Policy Committee on 19 March and 6 May 1975 and reports were given to the Committee of Council[10] on 15 May and 12 June of the same year.

For the success of a first-class laboratory it is from time to time necessary to get involved in ambitious projects carrying a considerable risk. This requires sufficient courage, but also a balanced judgement. CERN felt sufficiently strong to carry out a project in Europe which at the seminar in New Orleans was considered as a candidate for a world project (Fig. 1.3). The construction of such a machine had been suggested rather early at CERN by a long-term study group [6–8]. Of course, it was understood from the beginning that such a unique facility should be open to scientists from all over the world. It is not clear who at CERN first suggested the construction of such a machine, which got the name Large Electron–Positron Collider (LEP). The idea seemed rather obvious since it was a direct continuation of the European line of colliders such as AdA and ADONE in Italy, ACO in France and DORIS and PETRA at DESY in Germany. Anyway, CERN took the initiative before the international study group established at the New Orleans meeting had produced any useful results.

---

[9] A summary was given by Director-General Willibald Jentschke at the Scientific Policy Committee meeting CERN/SPC/371, 6 May 1975.

[10] Committee of Council CERN/CC/1183 12 June 1975. In the Committee of Council the delegates meet in closed session to prepare the Council decisions in open meetings.

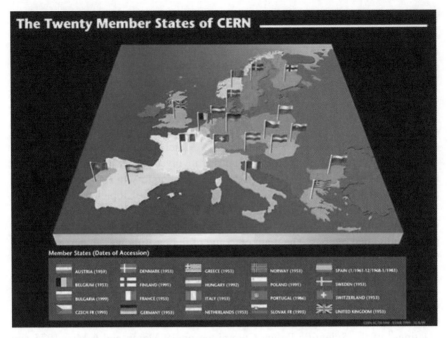

**Fig. 1.3** The present member states of CERN

## 1.4 The Birth of LEP

The first studies for such a collider started at CERN in 1976 in parallel with deliberations for other facilities, such as a large proton–proton collider as a follow-up to the very successful ISR and the electron–proton collider CHEEP [9][11]. CERN hesitated somewhat to leave the domain of proton machines, with which it had excelled so well, and gave in only slowly to the pressure of the physics community to look into electron machines. Indeed, ECFA recommended the construction of an electron–positron collider with an energy of at least 100 GeV per beam in 1977.

There were strong arguments in favour of such a facility which would open a new domain for the exploration of the microcosm. In the 1960s and 1970s a so-called standard model of elementary particles had been developed on the basis of a considerable number of experimental observations (see Chap. 8). Although this model was very successful in summarizing the experimental evidence on the building blocks of matter and the forces acting between them in a theoretical framework, it left many fundamental questions unanswered. A 'real theory' is based on a small number of basic assumptions and it should be possible to deduce the whole variety of observed

---

[11] This option was pushed by Björn Wiik, who later realized it at DESY as HERA. Unfortunately he died tragically much too early in an accident. The optimization of a large electron–positron machine was considered by Burt Richter during a sabbatical year at CERN in 1975.

effects by logical deductions ('reductionism') from them. In addition, predictions for new observations should be made to test the theory. The standard model in contrast starts from a relatively large number of assumptions and requires the input of several arbitrary parameters. It is based on the existence of six quarks and six leptons as building blocks of matter. Between these constituents of matter, three forces act which are transmitted by carrier particles (see Chap. 8). In particular, the predicted carriers of the weak force, the Z and W particles, and the top quark had not been discovered in 1976. The accuracy of the experiments was limited and hence a more precise confirmation of the existing data was needed and some additional predictions had to be tested.

The arguments why LEP could give answers to such and many other questions were summarized in 1976 in an excellent report by John Ellis and Mary Gaillard [10] which became the basis for the subsequent discussions. The electron–positron collider PETRA at DESY, Hamburg, Germany, started operation in July 1978 with an energy of 19 GeV per beam and a similar collider, the Positron–Electron Project (PEP) at Stanford Linear Accelerator Center (SLAC), Stanford, USA, started shortly afterwards, and various discoveries could be expected from those facilities. Nevertheless the physics arguments for LEP remained clear and valid. In a document [11] in 1979 it was still stated that "the situation might well be compared with making a spaceship journey to some other galaxy from which one has received already some faint signals." Assuming that the Z and W particles would be discovered by the antiproton–proton collider at CERN before LEP had come into operation (which indeed happened in 1983), an enormous amount of work remained to be done to investigate these particles in detail and to understand the various forces. Some burning questions were, for example, how many lepton families exist, is there a top quark and what is its mass, and is there a Higgs boson, a crucial element of the standard model whose existence is uncertain and whose mass cannot be predicted. However, it was also expected that important new information could be obtained for the strong interaction, e.g. does its strength change with the interaction energy, what are the properties of the gluons, are quarks and gluons definitely confined inside 'normal' particles or can they be liberated. Of course, one was also hoping for surprises.

In the end the outcome was similar to that for other facilities penetrating unexplored territory: some expectations were fulfilled, others not and some results were unexpected. This will be discussed in more detail in Chap. 8. The main attraction of an electron–positron collider lies in the fact that the events that occur in the collisions are very clean compared with proton–proton events and therefore can be relatively easily analysed and interpreted. The expectation that the LEP experiments would achieve a new level of precision was fully satisfied.

LEP was the electron–positron collider with the highest collision energy ever achieved. After 12 years of operation, it was replaced by the LHC, installed in the LEP tunnel, which will be the proton–proton collider with the highest energies reached in the world. Thus, the two most powerful facilities to explore the microcosm were realized at CERN, an extraordinary European success.

# References

1. Schopper H (2001) LEP, LHC and European perspectives on the SSC. Invited talk at the session 'Perspectives on major high energy physics projects', APS Meeting, 29 April 2001, Washington
2. Schopper H (1989) Materie und Antimaterie. Piper, Munich
3. Sessler A, Wilson E (2007) Engines of discovery. World Scientific, River Edge
4. Hermannn A, Krige J, Mersits U, Pestre D (1987) History of CERN. North Holland, Amsterdam
5. Lock O (1975) A history of the collaboration between the European Organization for Nuclear Research (CERN) and the Joint Institute for Nuclear Research (JINR), and with Soviet research institutes in the USSR 1955–1970, CERN/PE/ED/UR/75/3265. CERN, Geneva
6. Keil E (1974) Perspectives of colliding beams, CERN ISR-TH/74-22. CERN, Geneva
7. Richter B (1976) Very high-energy electron-positron colliding beams for the study of the weak interactions, CERN/IST-LTD/76-9. CERN, Geneva
8. Johnson K et al (1976) Design concept for A 100 GeV e/sup +/e/sup -/ storage ring (LEP), CERN 76-18. CERN, Geneva
9. CHEEP Study Group (1978) CHEEP: an e-p facility in the SPS, CERN 78-02. CERN, Geneva
10. Ellis J, Gaillard MK (1976) Physics with very high energy e+ e− colliding beams. CERN yellow report 76-18. CERN, Geneva
11. Dalitz R, Telegdi V (1979) Draft paper on the justification for LEP. CERN/SPC/435, 9 April 1979. CERN, Geneva

# Chapter 2
# The Difficult Decision of LEP's Size and Energy

When the decision has been taken as to which type of facility has to be realized, many questions concerning the details of its concept remain to be decided. For a machine designed for fundamental research, the following issues have to be taken into account:

- Will the machine be able to provide answers to the scientific questions raised by the scientific community in order to penetrate deeper into unexplored territory?
- Are the technologies available to realize such a facility or do new technologies have to be developed?
- Are the necessary scientific and technical staff available with the appropriate experience and competence?
- Last but not least, are there good chances that the necessary financial resources may be found?

Before a project can be proposed in a definite form, many discussions between scientists, engineers and politicians are needed. The final aim is, of course, to obtain a facility with the best performance at minimal cost. Since LEP was the largest device ever built, the answers to these questions were particularly pertinent and the choice of its parameters (size, energy) was important and was the most pressing decision to be taken.

## 2.1 The Optimization of Construction Cost

There is a fundamental difference for proton and electron storage rings. For a proton machine the highest achievable energy is determined by the bending power of the magnets which keep the particles in a circular orbit; hence, the most powerful proton rings use superconducting magnets which provide high magnetic fields. The conditions are different in the case of electrons. Electrons in a circular orbit emit (owing to the centripetal acceleration) a type of electromagnetic radiation, called

H. Schopper, *LEP – The Lord of the Collider Rings at CERN 1980–2000*,
DOI 10.1007/978-3-540-89301-1_2, © Springer-Verlag Berlin Heidelberg 2009

'synchrotron radiation'.[1] The energy loss $E_{rad}$ due to this radiation is given by the relation $E_{rad} \sim (E/m)^4/R$, where $E$ is the energy of the circulating electrons, $m$ is their mass[2] and $R$ is the radius of the orbit. This implies that the radiation losses increase very sharply with the energy $E$, whereas increasing the radius has relatively little effect. One has to fight against the fourth power of the electron energy with a linear dependence on the radius.

The radiation losses must be continuously compensated for by radio-frequency (rf) accelerating cavities which are fed by rf power supplies. If superconducting cavities are employed, the losses in the cavities can be neglected and for a given radius the rf power $P_{rf}$ has to increase with energy as $P_{rf} \sim E^4$. This was the case for LEP 2. If copper cavities are used, as was the case for LEP 1, there are additional losses in the walls of such cavities which increase with $E^2$ and consequently one obtains the overall relation $P_{rf} \sim E^8$. In both cases the rf power $P_{rf}$ must increase very steeply with the maximum electron energy for a fixed radius. This is costly both for the construction of the accelerating cavities and for the operation, which requires considerable electric power. The total construction cost is composed essentially of two elements: the cost of components, which is proportional to the circumference of the machine and hence to its radius (tunnel, magnet ring), and the cost of the rf system, which scales as $1/R$, as shown above. Optimization of the construction cost with respect to the radius shows that in the case of superconducting cavities both the radius and the construction cost increase with $E^2$, i.e. the radius of the machine should approximately increase with the square of the maximum energy. However, this relation is only very approximate and to find the most efficient parameters for an optimized electron–positron collider for a particular physics programme is not easy and requires compromises depending on the local and actual conditions.

The arguments given above apply, of course, only to a circular machine. If two linear accelerators with opposing beams are used instead, the synchrotron radiation losses are negligible and the length is roughly proportional to the desired energy (a fixed cost has to be added which is practically independent of the energy). The relations are shown schematically in Fig. 2.1 taking into account realistic prices for the various parts of the facilities.

It turns out that at a certain energy there is a crossover of the construction cost for the two types of machine, with linear accelerators being more economical. This crossover occurs at different energies depending on whether copper accelerating cavities or superconducting ones are used in the circular machine. Below a design energy of about 300 GeV a circular machine is more economical, whereas for higher energies two colliding linear accelerators are more advantageous. For

---

[1] This kind of radiation was observed in 1946 for the first time at a particle accelerator called a 'synchrotron' and hence its name [1]. Nowadays special storage rings are built to exploit this radiation for research and technical applications. It also plays a major role in astrophysics.

[2] The total power emitted by synchrotron radiation is proportional to $1/m^4$. Since protons are about 2,000 times heavier than electrons, the synchrotron radiation of protons is much weaker (about $10^{-13}$) and becomes noticeable only at extremely high energies, such as those at the LHC.

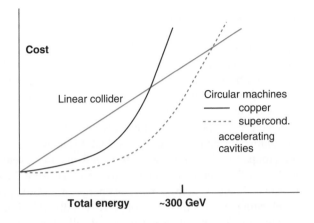

**Fig. 2.1** The construction cost of circular and linear electron–positron colliders as a function of the maximum energy of the facility. For the circular machines, the cases for copper and superconducting accelerating cavities are shown

a machine with only 3 times the energy of LEP the radius of a circular machine becomes unrealistic (several hundred kilometres in circumference). Since the design energy of LEP was well below the crossover of about 300 GeV, only a circular machine had to be considered, no matter what kind of accelerating cavity was considered.

On the other hand, for the next step in the development of electron–positron colliders only two colliding linear accelerators are feasible. Such a machine is presently considered as a world machine for a beam energy of more than 500 GeV with a total length of 30–40 km, the International Linear Collider (ILC) [2]. The disadvantage of linear colliders is that the particles in two opposite beams have only a single chance of colliding, whereas in a circular machine they can circulate many times, thus increasing enormously the chance of collisions. To obtain a sufficient number of collisions in a linear collider, very high beam currents (implying a large power consumption) are necessary and in addition the beams must be strongly focussed at the collision point. These requirements present serious technical and economic problems. To solve them further technical developments are necessary, and these are happening in a concerted way on a worldwide scale. A decision will only be taken after the results of these activities have come to fruition and when the results from the LHC are known.

Considering the arguments mentioned, it is not surprising that it took some time to agree on a final energy for LEP; the values suggested for energies oscillated up and down, leading to different circumferences for the ring. A first study group [3, 4] was formed at CERN in 1976 to examine the feasibility of a large electron–positron collider with a beam energy of 100 GeV (hence a total energy of $2 \times 100$ GeV) and a circumference of 50 km which was called LEP 100. When the technical study was terminated around the middle of 1977, several basic problems remained unsolved and the cost was considered to be very high. A new study had to be made.

## 2.2 The LEP Studies

In August 1978 the LEP Study Group issued a new design, LEP 70 (the 'Blue Book' [5]), for a smaller collider ring with a circumference of 22 km which could eventually reach beam energies of about 70 GeV. In a second stage the energy could be increased by replacing the accelerating copper cavities by superconducting ones if and when these became available. Since CERN always attaches great importance to having the support of future users, this design was presented to the European Committee for Future Accelerators (ECFA) and was discussed in ECFA-LEP working groups [6, 7] in September 1978, in Rome in November 1978 and in Hamburg in April 1979. The main conclusions were that from the physics point of view the machine should be able to reach an energy of 85 GeV per beam with conventional accelerating cavities made of copper. Secondly, it was stated that the underground area for at least some of the experimental halls should be larger than planned. Hence, the Blue Book design was not accepted.

The physics arguments were quite obvious. In a first stage one wanted a machine for the copious production of the still hypothetical Z particle ('Z factory'), which required about 50 GeV per beam, and in a second stage the threshold for W particle pair-production, estimated at that time to be 86 GeV, was envisaged. There was agreement that LEP would be needed even if the Z boson were discovered at the proton–antiproton collider at CERN (which happened in 1983) and if the top quark were found at PETRA in Hamburg (which did not happen). Many new results could be expected, e.g. information on the number of neutrino types, the discovery of the Higgs particle and new results on the strong interaction. Everybody agreed that such a machine would be a fascinating facility with no competitor in the whole world for a long time to come.

As a result of the in-depth discussions with the users' community, one more design was presented by the LEP Study Group (led jointly by Eberhard Keil, Wolfgang Schnell and Cees Zilverschoon) in the summer of 1979, the 'Pink Book' [8]. It was emphasized that this document was to be considered as a progress report and that the studies continued. The circumference of the machine was chosen to be 30.6 km and three energy stages were considered:

1. An initial 1/3 (zero) stage with a beam energy of 62 GeV
2. Stage 1, with an energy of 86 GeV per beam which could be reached with copper accelerating cavities
3. Stage 2, raising the energy to about 130 GeV by using superconducting accelerating cavities

In this proposal a special preaccelerator (injection) system was envisaged with two linear accelerators, an accumulator ring for the positrons and a synchrotron with a circumference of 1,741 m to be placed in a new tunnel under the Intersecting Storage Rings (ISR) site and using the ISR magnets (see Fig. 2.2 and Table 2.1).

One of the reasons why CERN originally was not considered as a good place for LEP was the fact that there was not much room between Lake Geneva and the Jura Mountains to place a tunnel with a circumference of more than 30 km. In the

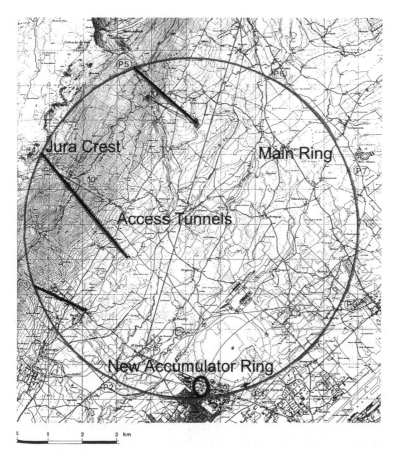

**Fig. 2.2** The LEP position in the first proposal ('Pink Book'). The ring passes deep under the Jura crest; three long access galleries were necessary to provide access to the underground halls at points *P3*, *P4* and *P5*. A new accumulator ring under the old Intersecting Storage Rings (ISR) tunnel was also proposed

**Table 2.1** The various proposals for LEP; energy with superconducting rf cavities in a second stage in *parentheses*

| Study | Maximum beam energy (single beam) (GeV) | Circumference (km) | Cost (millions of Swiss francs) | Year |
|---|---|---|---|---|
| LEP 100 | 100 | 50 | Too high | 1976 |
| Blue Book | 70 | 22 | ? | 1978 |
| Pink Book | 86 (120) | 30.6 | 1,300 | 1979 |
| Green Book | 50 (100) | 26.7 | 910 | 1981 |

Pink Book it was proposed to place it in such a way that about 12 km would be located in the rocks of the Jura, indeed passing under the crest at a depth of 860 m. However, very little was known about the geological features of the Jura. Some general information came from some water and road tunnels built in the vicinity of the Geneva basin. The tunnel would partly be located in a kind of sandstone, called 'molasse', between Lake Geneva and the foot of the Jura and limestone and other rocks under the Jura. The preliminary conclusion was [9]:

> These studies indicate that it is probably not too difficult to construct the LEP tunnel into the Jura with a boring machine. But to improve our knowledge of the molasse/limestone contact and to verify the quality of the Mesozoic limestone, it will be necessary to bore two reconnaissance tunnels of about 0.5 to 1 km long as soon as the location of LEP is determined.

As it later turned out, this expectation was drastically optimistic, the excavation of the tunnel becoming one of the main problems in the construction of LEP! The real weak point of the proposal was that the geological risks under the Jura could not be evaluated.

Eight experimental halls (see Fig. 2.2) of different types were foreseen: two 'surface' halls to be excavated from the surface, three underground halls up to 60 m deep and three deep underground halls up to 860 m deep. To reach these three deep underground halls it would have been necessary to dig three access tunnels 1, 1.5 and 2.3 km long, respectively, sloping towards the main tunnel with rather steep gradients. In this proposal a special preaccelerator (injection) system was envisaged with two linear accelerators, an accumulator ring for the positrons and a synchrotron with a circumference of 1,741 m to be placed in a new tunnel under the ISR site and using the ISR magnets (see Fig. 2.2).

Interestingly, a later possibility of colliding electrons from LEP with protons from the Super Proton Synchrotron (SPS) by installing an external 'bypass' for the protons from the SPS was also discussed. Therefore, LEP was placed in such a way that it passed close to the SPS. This possibility was never considered again.

The cost for stage 1 was estimated at CHF 1,275.4 million and the construction time considered was 7 years. It was the first time that a facility had been proposed to Council without a special budget for the project. In the past, special funds had always been approved for each of the new facilities (see Chap. 10). It was stated that LEP should be realized within a constant CERN budget and, of course, an austerity scenario had to be envisaged.

In parallel to the technical studies, the physics arguments in favour of LEP were presented to the Scientific Policy Committee (SPC) of CERN. A report by Richard Dalitz and Valentin Telegdi [10] which was based on the results of the users meeting at Les Houches in the French Alps mentioned above was discussed in an SPC meeting in April 1979. A more formal document [11] was presented and adopted at the SPC meeting of 18–19 June 1979. The reasons for choosing an electron–positron collider were given, the physics case for the energy chosen was explained in detail and, in particular, the various stages for the increase of the energy were considered. The feasibility of constructing such a machine was pointed out, with the somewhat naive argument "since the cost of the tunnel is but a small fraction of the capital cost, its circumference should certainly be made large enough for acceleration to the highest energies discussed here". Indeed it turned out in the end that the tunnel was the most costly and difficult part of LEP!

Contrary to the initial ideas, it was suggested in the Pink Book that LEP be built at the CERN site although "it is not yet definitely clear whether this is technically feasible, but it would in any case avoid all kinds of difficulties and delays commonly associated with site selection."[3] It was stated that it was imperative to start constructive actions immediately and no time should be wasted and Council was asked for approval to be achieved by the end of 1981.

# References

1. Baldwin GC (1975) Phys Today 28:9
2. Barish B, Walker N, Yamamoto H (2008) Building the next-generation collider. Sci Am 298:46
3. Camilleri L et al (1976) CERN 76-18. CERN, Geneva
4. Bennett JRJ et al (1977) CERN 77-14. CERN, Geneva
5. The LEP Study Group (1978) The blue book. CERN-ISR-LEP/78-17. CERN, Geneva
6. Jacob M (ed) and CERN/SPC (1978) Proceedings of LEP summer study, ECFA-CERN 79-01, Les Houches and CERN, 10–22 September 1978. CERN, Geneva
7. Llewellyn-Smith C (1979) Physics' Justification for LEP, CERN/SPC/435, appendix B. CERN, Geneva
8. LEP Study Group (1979) Design study of a 22 to 130 GeV e+e− colliding beam machine (LEP). The pink book (1979). CERN-ISR-LEP/79-33. CERN, Geneva
9. LEP Study Group (1979) CERN-ISR-LEP/79-33. CERN, Geneva, chap 14, p 164
10. Dalitz R, Telegdi V (1979) Draft paper on the physics justification for LEP. CERN/SPC/435, 9 April 1979. CERN, Geneva
11. CERN (1979) Physics justification for LEP. CERN/SPC/435/Rev, 15 May 1979. CERN, Geneva

---

[3] When a site for the SPS in another member state was discussed in the early 1970s, no agreement could be reached and in the end the SPS was built at the Geneva site.

# Chapter 3
# The Approval, or How To Persuade Governments

For the final approval of the LEP project various obstacles – political, financial, technical and geological – had to be overcome. The 'golden times' of CERN during which the CERN budget increased almost exponentially for several years (see Chap. 10) belonged to the past and the Super Proton Synchrotron (SPS) project was the last one for which special funds had been allocated. From 1981 on, CERN had to live with a constant budget. This implied tough consequences for the management of the laboratory and its users. From the point of view of the history of science policy it may be interesting to go through the different steps of the LEP approval procedure.

## 3.1 The Unification of CERN

The 'Pink Book' (see Chap. 2) was presented to the subcommittees of the CERN Council in the second half of 1978 and the whole of 1979 was spent in discussions on whether and how LEP could be realized. At that time Council started to discuss a reorganization of CERN and to look for a new director-general. At the time when the construction of the SPS machine had been considered at the beginning of the 1970s, the member states intended to open a new site and CERN had been separated into two laboratories, CERN I and CERN II. However, after a long procedure during which new sites in all 12 member states were investigated, it turned out that no agreement on a new site could be reached. However, John Adams, who had been crucial in the construction of the Proton Synchrotron (PS) accelerator, had prematurely been appointed as director-general for the new laboratory CERN II. To stop the quarrel about the new site, Adams proposed building the SPS at the existing site[1] in Geneva, and this proposition was finally adopted by Council. In 1970 Willibald Jentschke was appointed as director-general for the existing laboratory CERN I, with the result that two laboratories existed side by side in Geneva. CERN II had the task of building the SPS, whereas CERN I continued the ongoing physics programme. When Jentschke's term ended in

---

[1] Following suggestions by Colin Ramm and Günther Plass elaborated in a long-term study group

H. Schopper, *LEP – The Lord of the Collider Rings at CERN 1980–2000*,
DOI 10.1007/978-3-540-89301-1_3, © Springer-Verlag Berlin Heidelberg 2009

1975, the two laboratories were formally reunited; however, two directors-general remained, Leon van Hove as research director-general and John Adams as executive director-general. CERN II became the SPS Division, but in reality this division continued as a rather independent unit. In retrospect, the decision to build the SPS at the Geneva site was essential for the long-term future of CERN. Only under this condition were the available resources, both human and financial, concentrated in one place and could be redeployed later for the construction of LEP and later the LHC.

Council decided that at the end of 1980 when the terms of van Hove and Adams came to an end the laboratory should be fully reunited under a single director-general. Normally a new director-general is nominated 1 year before his mandate starts. Therefore, on 19 December 1979 the Committee of Council,[2] whose meetings are not public, agreed on me as the only candidate, with 11 votes out of 12. The main reason for selecting me was probably that I had been chairman of the Directorate of DESY and hence had ample experience of large projects and in particular with electron–positron colliders, having been responsible for the construction of PETRA. However, the delegate of one country (Italy) had instructions not to cast a vote. Apparently Italy suspected that my heart would still be beating for DESY and that I would not fight for LEP at Geneva, even while being director-general of CERN. Italy also considered a prominent Italian scientist as a possible candidate. Since traditionally the directors-general are elected by unanimity, the Committee of Council at its session on 19 December 1979 did not take a formal decision in order to allow Italy to change its position and join the majority.

However, time was pressing and a decision had to be taken at the Committee of Council meeting scheduled for 29 February 1980. Indeed a decision was taken. In a CERN press release [1] it was said that "at this meeting, and in particular at the request of the Italian delegation, the future programmes of the Organization were discussed, and the proposed new European accelerator project LEP and the calendar of decisions related to it were examined." It was further communicated that:

> At the same meeting the Committee of Council took note that the twelve delegations unanimously supported the appointment for five years of Professor Herwig Schopper to the post of Director-General of the Organization, from 1st January 1981.... The Committee of Council entrusted Professor Schopper with the mandate to present to Council, at its session in June 1980, his proposals concerning the top management structure and the directing staff of CERN.

So everything had gone without problems? Not quite! What had happened behind the scenes was the following. In the morning before the Committee of Council session, I had a private meeting at breakfast with the Italian delegate in the rather

---

[2] The Committee of Council consists of delegates of member states, but not of observers and meets between Council sessions in closed session and prepares the decisions of Council, which meets in public session.

cheap hotel where I was staying. I managed to convince him that I would do my best to get LEP approved by CERN. Antonino Zichichi, an eminent Italian staff member of CERN, was also present and although he had been considered by Italy as a candidate, he helped in a fair way to convince the Italian delegate, Umberto Vattani. However, to convince him personally was not sufficient, he had to get new instructions from Rome. Therefore, we agreed on the following plot. I would be invited to give a report to the Committee of Council explaining my concept of LEP, then during the subsequent coffee break the Italian delegate would call Rome to ask for and receive new instructions and finally would be able to vote in my favour. That is how it happened – but, of course, not a word of this is recorded in any document! The whole operation was guided very skilfully by Jean Teillac (Chairman of the French Atomic Energy Commission CEA), who even had the courage to send me a letter of appointment dated 26 February 1980, which made it possible to look immediately for my successor at DESY; Volker Soergel was appointed. My nomination was approved formally and happily in a special Council meeting on 25 April 1980, on which occasion the Italian ambassador expressed warm congratulations.

In the following months a very efficient and agreeable cooperation developed between Adams, van Hove and me. We prepared a new design proposal which was presented to Council at its meeting in June 1980. It was still based on the Pink Book, with a circumference of 30 km, but with one major modification. Instead of foreseeing a new preaccelerator system, the group responsible for the injection system had proposed using the SPS and the PS as preaccelerators. This brought down the cost of the project to CHF 950 million, a level which we considered tolerable. If the project were approved in 1981, preliminary operation could start in 1986.

The use of the SPS and PS as injectors for LEP had not only technical consequences but made it necessary to change the formal frame of the LEP project. Since the SPS and PS remained the main facilities for the normal CERN programme, LEP could no longer be considered as an independent project but had to become part of the overall 'Basic Programme'. A consensus on the procedure for the approval of LEP by Council was established in its session in December 1980 [2]. An essential sentence read: "If the inclusion of the LEP project. . .in the Scientific Activities and Long-Term Budget Estimates is agreed by the Council *with no Member State voting against*, this will constitute approval of the LEP Project." Inclusion of LEP in the Basic Programme of CERN was finally decided at a Council meeting [3] in May 1981. This clearly meant that LEP could not be realized by only some of the member states, but all member states had to participate and a unanimous decision (with possible abstentions) was required. It also implied the inclusion of LEP in the so-called Bannier procedure, which was a rolling annual budget estimate for a period of 5 years, and once a budget ceiling had been adopted, it could be increased only "if no objection raised", whereas for a reduction of the ceiling a two-thirds majority was sufficient. I was asked by Council to present a definite proposal in June 1981 and a financing scheme providing the integration of LEP into the Basic Programme *with no additional funds*.

## 3.2  Adapting to the Austere Conditions

To obtain the required unanimous decision, I travelled at the beginning of 1981 to most capitals of CERN member states to convince the politicians of the soundness of our project and to sound out under what specific conditions they would be prepared to approve the LEP project. Perhaps it was not surprising that most of the opposition came from colleagues in other fields. They feared that the construction of LEP within a constant budget was unrealistic and that once it had been approved CERN would come back with additional requests which would be difficult to refuse. This could have negative effects on the financing of other scientific fields in the member states. These worries were particularly expressed in countries where the CERN contribution was funded from the normal science budget. Since increases of the total science budgets by parliaments were not very likely, an increase of the CERN contribution automatically meant a reduction of the funding for other projects. What made the problem even more acute was the fact that the CERN contributions had to be made in Swiss francs, a currency which was very strong at those times, and fluctuations in the exchange rate could therefore have devastating effects. I recognized that all efforts would have to be made to avoid such a development and that LEP would have to be built strictly within the given financial boundaries in order not to damage the credibility of CERN.

To build LEP with the resources available within CERN it seemed necessary to introduce a new organizational system to exploit them in the most efficient way. In the past, the organization of CERN was based on divisions for accelerators, physics and services. Over the years some changes had been introduced mainly for the purpose of adapting the structure to new projects or to the preferences of directors-general. These divisions were directed by extremely competent scientists or engineers, with strong personalities, with the consequence that they became rather independent. As a result, the optimization of resources had been applied to individual divisions rather than to the overall situation of CERN. With the separation of the laboratory into CERN I and CERN II, this tendency became even more pronounced. During the 'golden times' when the CERN budget increased rapidly (see Chap. 10) this kind of policy seemed to be justified and successful. To adapt to the new conditions it was absolutely necessary to break down to a certain extent the borders between divisions or at least to make them more permeable. This was not only a financial problem but above all it was a harsh human problem. The strength of CERN was based, apart from the scientific and technical competence of the staff, on the excellent human relations between all levels of the hierarchy which had developed over several decades. Some of these links now had to be ruptured and this concerned not only a small fraction of the staff, but indeed it turned out that about one third of the staff had to be transferred to new tasks, implying a remarkable mobility inside the organization.

I had to take these aspects into account when it came time to appoint a LEP project leader. Among the accelerator people there were some who had been responsible for earlier projects, had played a decisive role in the preparation of the early LEP proposals and had all the qualities to promise full success for the LEP project. Hence, many were quite surprised or even disappointed when I asked the Italian Emilio Picasso to lead the LEP project. He was an experimental high-energy physicist and not an accelerator specialist. However, during his career he did experiments which required close collaboration with the technical divisions and therefore he enjoyed enormous respect and recognition not only among the physicists but also among the accelerator engineers. In addition, he had the temperament and the personality to smooth out human problems. It turned out that this was an excellent decision. Emilio Picasso did a superb job. I enjoyed my close cooperation with him over more than 8 years and we became friends, a fact which became essential when we had to take very delicate decisions together by ourselves, as I shall explain later. A new division was created for the project, the LEP Division, and Günther Plass, an experienced accelerator expert and former deputy PS division leader, was nominated as division leader and deputy to Picasso.

## 3.3  The Final Proposal – with the LHC in Mind

With the new and strict conditions established by the member states, we had to prepare a new proposal for LEP. Our first important task was to fix the main parameters of LEP and reconsider its location. For more than 2 years test borings had continued to explore underground. The underground geological features of the region could be distinguished in two parts. In the plain between the foot of the Jura and the airport, a special kind of sandstone (the molasse) was quite good for tunnelling, whereas the limestone under the Jura was very bad. It appeared particularly important to study the critical transition between the two kinds of rock and a reconnaissance tunnel was being excavated. One result was that the long access galleries (Fig. 2.2) foreseen to reach some of the experimental underground halls would have to pass through very bad parts of rock. The real great danger, however, was the 12 km of tunnel under the Jura with rock of calcareous nature with big caverns and possible infiltrations of water under high pressure and a covering layer up to almost 1,000 m, which in the case of an accident made access from the surface impossible. In addition, deep down in the Jura there was rock from the Triassic Period which was considered very dangerous for tunnelling because of its plasticity (see Fig. 3.1). When Emilio Picasso and Henry Laporte, who was appointed to direct the construction of the tunnel, met Prof. Giovanni Lombardi, a worldwide expert in tunnelling, they were asked how much money was available for the tunnelling. When Lombardi learned about the restricted budget he replied: "You either get the tunnel out of the mountain as much as possible, or my advice is to let others build this tunnel."

After long and sometimes heated discussions we decided to reduce the circumference of LEP from 30 km to about 27 km and move it somewhat out of the Jura (Fig. 3.2) to reduce the geological risks, but 8 km of tunnel was still under the Jura. Fortunately, the clever design group of LEP found a more favourable tuning for the beam dynamics which allowed the design energy of phase 1 (50 GeV per beam) to be reached even with the reduced tunnel circumference.

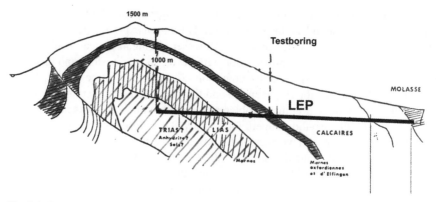

**Fig. 3.1** Geological cross-section showing the position of the LEP tunnel according to the original proposal. A section would be deep under the Jura Mountains in bad rock

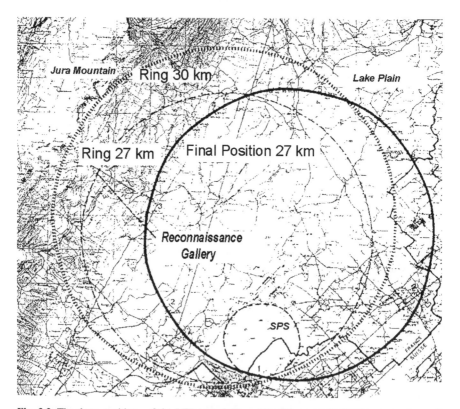

**Fig. 3.2** The three positions of the LEP tunnel: the originally proposed (30-km circumference), the intermediate choice as approved by Council (27 km) and the final position

When this decision became known, I was heavily criticized by many colleagues and two prominent ones expressed their worries even in writing. John Adams, one of the most esteemed accelerator experts in the world, sent me a confidential note (dated 12 March 1981) saying "it seems to me that your choice now is either to battle on with the 27 km circumference LEP with possible delays in starting construction, continuous trouble with the French authorities at all levels and a serious risk of delays and overspending on the project, or to go flat out for a smaller LEP this year which would avoid all these problems." In summarizing, he proposed to reduce the LEP circumference to 22 km. On 6 April 1981 I received another confidential note, from Carlo Rubbia, who was already an eminent physicist even before he received the Nobel Prize in 1983. He wrote:

> I am convinced that your recent decision to reduce the machine radius has been very wise to the extent that it removes some fraction of the traversal of the Jura. I believe however that one should go further and avoid the mountain completely. This corresponds approximately to a new circumference of the order of 23 km. . . . To conclude: penetrating the Jura may open the way to a major geological disaster. Since there is no way of being a priori sure that this will not happen, I would strongly advocate that one takes the fastest and safest solution of remaining under flat land. . . .

The opinion of these two respected colleagues caused me some headaches and I had a few bad nights. Finally I decided to maintain the tunnel size as planned at 27-km circumference. The main reasons for accepting the risks attached to such a ring circumference were the following. Reducing the size would certainly not have impaired the performance of LEP considerably. However, I was aware that in the distant future many changes could be imagined, even removing the whole machine from the tunnel, whereas the only item which could not be changed was the size of the tunnel. In view of the long-term future of CERN and of a later project, a proton collider in the LEP tunnel, a possibility which had been considered already at that time, I thought it would be decisive to have as large a tunnel as possible. Fortunately fate was kind to us and the big disaster did not happen. On the other hand, the existing LEP tunnel was certainly one of the major positive elements when the LHC project was approved and its large circumference made it possible to reach interesting energies.

In June 1981 I submitted the final proposal, the 'Green Book' [4], to Council with a detailed cost estimate. To have a chance of receiving a positive decision, a new approach had to be adopted which became known as the '*stripped-down LEP*'. It consisted of the following main elements:

1. LEP was presented as an evolving machine, implying that only the absolute minimum number of components would be installed for phase 1, just sufficient to produce Z particles in abundance ('Z factory'). The upgrading would have to be decided on later.
2. In spite of these savings, the future potentialities should not be impaired. In particular, the tunnel circumference should be made as large as possible, with no reduction of the tunnel cross-section in view of a later proton ring in the LEP tunnel. However, we decided to reduce the tunnel circumference from 30 km

to about 27 km and move it somewhat out of the Jura so that only about 8 km (instead of 12.5 km) would be in the bad limestone. The access galleries could also be shortened accordingly (see Fig. 3.2, ring in the middle).

3. Only four instead of eight interaction points would be equipped as experimental areas.

4. The project would carried out with a constant budget. It would be the first time at CERN that no additional funds would be required for a project. According to the budget rules to be adopted for the LEP project, an increase of the constant budget level required unanimity, whereas a decrease could be decided by a two-thirds majority.

5. The total investment for the project was reduced to CHF 910 million (in 1981 prices) from the approximately CHF 1,300 million of the Pink Book. This amount included CHF 20 million for the experimental infrastructure. According to CERN practice, the cost estimate did not include any salaries for CERN staff, but the existing personnel without additional posts would have to build LEP. A financing period of 8 years was foreseen from the date of approval, which implied that some bills could be paid only after the machine started operating ('debts'). However, every effort was promised to complete the machine as early as technically possible.

6. No contingencies were foreseen in the budget, with the argument that the main uncertainty was in the civil engineering of the tunnel and the associated financial risks were almost unpredictable in view of the difficult geological situation. Any major problem in tunnelling would unavoidably lead to delays and hence the financing period would be automatically extended. '*Time is the contingency*' was our slogan!

7. No provision was made for the experiments since it was not known yet how many experiments could be approved and what the amount of outside contribution would be.

8. I proposed three alternative levels for the constant budget which would allow LEP to be built and the other programmes to be continued at a reduced but still acceptable level.

Of course, these conditions did not correspond to what is usually necessary for a big project. In particular, points 6 and 7 were abnormal. Comparing the cost estimate of LEP with that of other projects, one should remember that no contingencies and no CERN staff costs were included.

## 3.4 The Painful Approval Procedure

An extremely severe problem concerned the level of the 'constant' budget mentioned in point 4 in the previous section. During the discussions three levels of the 'constant budget' during the construction period were considered: levels A, B and C, with CHF 629, 650 and 671 million, respectively. Since it was obvious that it would be very difficult to get a positive decision at all, I asked for level A. The Scientific Policy Committee considered such a level as barely acceptable and a higher level was strongly supported. Very tough discussions in the Finance Committee and the Committee of Council followed and finally in the Council session in June 1981 a decision was taken. The member states were split – some were prepared to agree to a figure of CHF 629 million, some did not want to give more than CHF 610 million and the final compromise was CHF 617 million – a level well below the one I had asked for. The difference with respect to the requested budget of CHF 629 million may seem not very dramatic, but the difference multiplied by the 8 years of construction gave a sum of CHF 96 million, i.e. about 10% of the total project cost.

I proposed the following gentlemen's agreement. The total CERN budget would be subdivided into a material and a personnel part, and these would be indexed separately for inflation. I requested that during the construction of LEP at least the full inflation index for the material part of the budget should be promised at least informally. However, the answer was, "You accept the budget level and we shall tell you each year how inflation will be compensated for." The result was that every year long discussions concerning the index took place. Difficulties will be explained later in Chap. 10.

With this budgetary decision I had to face the options of trying to build LEP under such harsh conditions or rejecting the decision. During the long preceding discussions I became convinced that the member states were not willing to make more funds available for CERN and this was in a way confirmed by the subsequent very wearisome procedure for the final approval, which had to be unanimous as explained above and which was at certain moments close to failure. It was obvious that success would require the mobilization of all resources inside CERN and an enthusiastic and understanding staff. I was criticized by many colleagues and in particular by the Staff Association[3] of CERN for accepting these conditions for the financing of LEP. The Staff Association was, of course, aware that in a time of extremely short funds the conditions for employment might be impaired and they reproached me for wanting to "build LEP at the cost of the staff". They asked me to resign. I was not surprised and was somewhat amused to hear some 20 years later that the same arguments for the financing of the LHC were put forward.

With these conditions LEP could be built only by introducing new, sometimes very painful measures:

1. The construction period had to be extended by 1 year until the end of 1988 in order to accommodate the expenditures in the constant budget.
2. The 'stripped-down' policy had to be applied even more strictly, providing only those machine components which were absolutely necessary to provide a first beam at 50 GeV.

---

[3] CERN as an international organization has no trade unions. The interests of the staff are defended by the Staff Association, which reports to the director-general but has no formal right to negotiate social issues directly with the Council.

3. The non-LEP programmes would have to be considerably reduced.
4. It would be necessary to transfer funds from the operation budget to investments.
5. The CERN contributions to the LEP experiments would have to be considerably lower than in the past, requiring more funds and personnel from outside users.
6. The services would have to be rationalized in every possible way, provoking unavoidable complaints from users.

Even with all these very restrictive conditions, the final approval for LEP could not be achieved in the Council session on 25–26 June 1981 presided over by Jean Teillac, who tried everything to get a positive outcome. Eight member states (Austria, Belgium, Germany, France, Greece, Italy, Switzerland and the UK) cast a positive vote, Denmark cast a positive vote only ad referendum (it was not completely clear what that meant!) and three countries (The Netherlands, Norway and Sweden) declared that their internal decision procedure was still going on. Hence, for a few months the trembling continued. A big sigh of relief could finally be uttered when at a special Council meeting on 30 October 1981 the unanimous approval was achieved.

## 3.5  The Thorny Consequences of a Limited Budget

A few remarks should be added to the conditions mentioned already. The extension of the construction period of LEP by 1 year was later occasionally misunderstood as a delay in the realization of the project, whereas it had already been decided at the time of the authorization. The target date of the end of 1988 meant, of course, that the first beam would be obtained in spring 1989, since during the winter CERN was usually shut down to save on electricity costs, which are higher in winter. Indeed, with the first collisions on 13 August 1989, a real delay of only a few months occurred, an excellent achievement in view of the geological difficulties to be described in Chap. 4.

That LEP could be built within a constant budget was possible thanks to the flexibility mentioned under point 4 in the previous section. In most national environments a transfer of funds from investment to operation or vice versa is not allowed, mainly because the two parts of the budget are approved separately by national parliaments since normally debts are allowed only up to the sums invested. When German Research Minister Heinz Riesenhuber visited CERN, I pointed out to him that only because CERN was not bound by such restrictions was it possible to build LEP within a constant budget. He became so convinced that he thought the same flexibility should be introduced at German research centres. However, he found out that this was completely impossible because of the reasons given above.

Another comment concerns the reduction of the non-LEP activities. The most exciting part was the proton–antiproton collider programme at the SPS, where the discovery of the W and Z particles of the weak interaction was hoped for. In the early 1980s it turned out that additional funds were necessary for improvements to the running experiments UA1 and UA2. However, extra funds had to be attributed in a somewhat informal and confidential way since ironically the upgrading of the SPS to a proton–antiproton collider, perhaps the most successful activity of CERN, had never been formally approved as a regular CERN project. This created additional problems for the financing of LEP, but certainly this extra money was very well spent with the discovery of the W and Z particles in 1983, for which Carlo Rubbia and Simon Van der Meer were awarded the Nobel Prize.

However, many other non-LEP programmes had to be terminated completely:

- The ISR, the only proton–proton collider in the world still doing excellent physics, had to be stopped in 1983, a particularly painful decision.
- The Big European Bubble Chamber (BEBC), whose construction had been financed with special contributions from France and Germany, had to be closed down.
- Almost the whole fixed target programme at the PS had to be slowly phased out.
- The SPS fixed target programme in the so-called west hall had to be considerably reduced.
- The Synchro-Cyclotron (SC) operation,[4] mainly devoted to very interesting nuclear physics with the isotope separator ISOLDE, had to be reduced from 6,000 to 4,000 h per year.
- The long-term accelerator research had to be limited to the development of superconducting accelerator cavities and magnets. Both these technologies were crucial for the upgrading of LEP and the construction of the LHC, respectively, but funds totalling no more than 1% of the budget could be allocated to them.

All these facilities were still producing excellent physics involving several hundred scientists. Hence, it was not a matter of closing down obsolete facilities, but tough and painful priority choices had to be taken by the Directorate, taking into account, of course, the advice of various external advisory committees. At that time I lost many friends among my fellow scientists, but fortunately I could later regain most of them!

In spite of all these restrictions a new programme was started during the LEP construction period, investigations with accelerated heavy nuclei. I decided in favour of this 'heavy ion' programme against the recommendations of the advisory committees, a rare case where a director-general had to use his full power! The financing came largely from extrabudgetary sources, but a heavy load remained for the CERN accelerator experts. It implied that heavy nuclei (e.g. lead nuclei) were accelerated in the PS and SPS to be fired at heavy metal targets. The long-term motivation was to find the theoretically predicted quark–gluon plasma which played an important role in the very early stages of cosmic development. This programme will be continued

---

[4] The SC was the first facility to be built at CERN, and had a proton energy of 600 MeV.

with the LHC, where the special large detector ALICE will be devoted to this field. ALICE is located in the cavern where the L3 experiment had been installed and it uses the huge magnet of L3.

## References

1. CERN (1980) Press release 'Réunion officielle au CERN, CERN/PR 8/80, 4 March 1980. CERN, Geneva
2. CERN Council Procedure for approving and starting the LEP project, (1980) CERN/1395, December 1980. CERN, Geneva
3. CERN Council Resolutions – approval of the LEP project, phase 1 – revision of the bannier procedure with a view to the LEP project, (1981) CERN/1411, 22 May 1981. CERN, Geneva
4. CERN (1981) The green book. CERN 2444. CERN, Geneva

# Chapter 4
# The Tunnelling Adventure

The realization of LEP presented many demanding technical difficulties, but the excavation and construction of the tunnel was the most difficult and hazardous challenge. If the tunnel construction had failed, the whole project would have been doomed, of course. The problems to be surmounted were the geological features, the environment, political issues and the choice of the final location. One particular difficulty inherent to the project was the requirement of a precise shape for the tunnel, consisting of eight circular arcs and eight straight sections in-between. Whenever in other tunnel projects a special location presented particular problems, it may have been possible to deform the tunnel in such a way as to bypass the problematic zone. In the case of LEP this possibility was excluded since the geometry of the ring had to be respected with a tolerance of centimetres to guarantee the proper behaviour of the particle beams. The total cost of the civil engineering turned out to be about one third of the total project cost. These circumstances determined in an essential way the project as a whole; therefore, I should like to describe in some detail the civil engineering and, in particular, the underground work. In the 1980s the LEP tunnelling was the largest construction project in Europe until the excavation of the tunnel under the English Channel between France and Great Britain.

LEP is situated between Lake Geneva and the Jura Mountains (Fig. 4.1), and would be one of the few human-built objects sufficiently large to be detectable from space if it were not underground. The region, mainly used for agriculture, is known for its natural beauty and it stretches on both sides of the border between Switzerland and France. The LEP tunnel crosses the border several times (Fig. 4.2). This implies that many people have to go from the main laboratory in Switzerland to sites in France and a considerable amount of material has to be transported back and forth. This has never created any serious problems, an impressive example of international cooperation and trust from the host states.

## 4.1 The Different Elements of Civil Engineering

The civil engineering of LEP as determined by the needs of the machine and the experiments was quite complex. One tends to think mainly of the tunnel itself. Paradoxically, however, less than half of the total rock which had to be

H. Schopper, *LEP – The Lord of the Collider Rings at CERN 1980–2000*,
DOI 10.1007/978-3-540-89301-1_4, © Springer-Verlag Berlin Heidelberg 2009

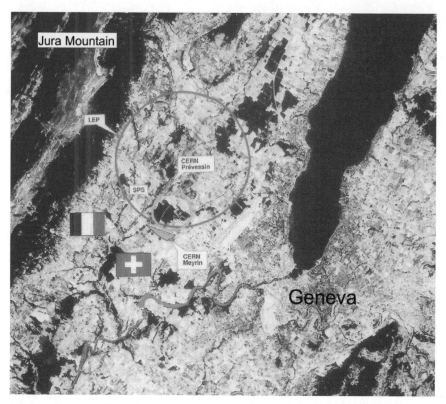

**Fig. 4.1** Satellite photo of the Geneva region. LEP is situated between Lake Geneva and the Jura Mountains

excavated (more than $1.4 \times 10^6 \, \text{m}^3$) came from the main tunnel. The various components were:

- The main tunnel consisting of eight circular arcs with eight straight sections in-between with a total circumference of 26.6 km. The diameter of the tunnel to be bored was 3.8 m, much smaller than a road tunnel. However, the precision concerning the shape of the tunnel was extraordinary (see Chap. 6).
- Access shafts to bring down equipment for the machine and the experiments, but also for the safety of the people. In total, 18 shafts had to be excavated, with a total length of several hundred metres, the deepest being 150 m (Fig. 4.3).
- Parallel to the straight sections additional galleries had to be excavated for the installation of auxiliary equipment, e.g. the klystrons to feed the accelerating cavities in straight sections 4.2 and 4.4.
- Four huge underground halls to house the experiments had to be created.
- In addition, 55,000 m$^2$ of buildings at the surface had to be constructed.

**Fig. 4.2** Air view of CERN. The LEP ring (now housing the LHC) crosses the Swiss–French border (indicated by a *dotted line*) several times. The previously built Super Proton Synchrotron (*SPS*) ring is shown as well as the site of the main laboratory – Proton Synchrotron/Intersecting Storage Rings (*PS/ISR*). In the foreground Geneva airport can be seen and the Jura Mountains are at the back

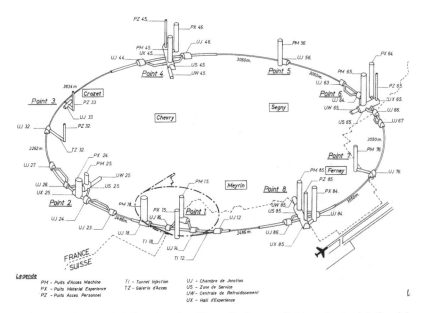

**Fig. 4.3** The underground civil engineering. The main ring, parallel tunnels at points 2 and 4 and the access shafts

## 4.2 The Geology and Hydrology

Originally it was thought that LEP could not be built near Geneva since the geological conditions of the ground were not known very well and seemed unfavourable. The plain between the Jura Mountains and Lake Geneva consists mainly of a kind of sandstone called 'molasse'. The 6 km of the Super Proton Synchrotron (SPS) tunnel was excavated practically only in this kind of rock. At that time this was considered a major adventure, although looking back it seems a simple task, with the only problem being that some gas pockets were encountered. However, for LEP several kilometres of the tunnel were supposed to pass under the Jura, which consists of limestone with many faults and cracks, partly filled with water under high pressure (because of the covering rock up to an elevation of 1,600 m). This made the construction of the tunnel a real adventure.

The first important task was to get as much knowledge about the geological formation and the hydrological features of the Geneva basin as possible in order to be able to evaluate the risks and draw conclusions for the tunnelling strategy. This work was carried out under the responsibility of Henry Laporte with the help of other staff (e.g. Bruno Bianchi, Manfred Buehler-Broglin) in collaboration with the competent French and Swiss authorities and external experts.

An exploration tunnel about 4 km long going from the plain to the foot of the Jura had been started earlier, mainly to explore the transition from the molasse in the plain to the limestone under the Jura (see Fig. 2.2). Since more information was needed, a programme of test borings was started in 1980 and 1981; the total length of the vertical boreholes was 9 km, including three deep borings in the Jura, one going to a depth of 1,000 m. They showed that deep under the crest of the Jura the rock was very bad, consisting of Triassic anhydrite salts inadequate for tunnelling (see Fig. 3.1). Higher up some karstic cavities were found, several metres in diameter. Geological faults which had been identified already from the surface were confirmed, in particular a major fault near the river Allondon. Water pressures up to 20 atm were detected.

In the plain the extension of the molasse, which is very good for tunnelling, could be established. However, it was found that towards the airport its roof went down and the coverage by moraines increased instead. Underground water reservoirs were also detected in some areas above the molasse and these would have to be traversed when the access shafts were excavated.

Concerning the hydrology, it was well known that some of the villages in the Pay de Gex, the French region at the foot of the Jura, suffered from lack of water in dry summers. Hence, good knowledge of the hydrological features was important, not only for the tunnelling strategy, but also to avoid an increased water problem for the population. Therefore, an intensive hydrological study was started in 1981, in close cooperation with an outside expert (Prof. Albéric Monjoie, University of Liège), and continued during the whole LEP construction period. Instruments to survey the water flow in several rivers draining the Jura and rain gauges were installed.

As a result of all the measures taken, the area around LEP is one of the best studied as far as geological and hydrological features are concerned [1].

## 4.3  The Choice of the Final Position

Choosing the final position of the LEP main ring was a delicate business. Not only were the consequences of the risks and the cost of tunnelling of paramount importance, the position of LEP was also drastically influencing the price of land, depending on whether it was close to or farther away from the tunnel access shafts, an issue of significance for the people living in the area. Therefore, we could not discuss the positioning publicly and only Emilio Picasso, Henry Laporte, Giorgio Brianti and I were involved. The possible positions were limited technically by the request that the main LEP tunnel should join the SPS tunnel more or less tangentially, since the existing CERN accelerators were to be used as preaccelerators.

As I explained in Chap. 2, the original proposal with a ring circumference of 30.6 km had to be given up for geological and cost reasons. In this original proposal the ring would have passed right below the Jura crest at a depth of almost 1,000 m and we had learned from the test borings that the rock there was unacceptable for tunnelling. In June 1981 we proposed the 'stripped-down' version of LEP with a circumference of 26.6 km and it offered the advantage of reducing the part of the tunnel under the Jura from 12 to 8 km (see Fig. 3.2). When I presented the proposal for LEP to Council in June 1981 this position seemed a reasonable compromise.

However, on 9 October 1981 Picasso sent me a confidential memorandum in Italian, the language he used when was very worried. The note accompanying the memorandum[1] demonstrated our excellent relations based on full mutual

**Fig. 4.4** LEP project leader Emilio Picasso (on the *right*) and the author

---

[1] It read "Caro Herwig, Questa mia nota é scritto non per farti infelice o renderti scontento. É per chiarire i nostri punti di vista. É indirizzata solo a te e a nessuno altro. Ed é scritto con molta stima per te. Emilio". In English: "Dear Herwig, This note of mine is not written to make you unhappy or dissatisfied. It is to clarify our viewpoints. It is addressed only to you and nobody else. And it is written with great respect for you. Emilio".

trust (Fig. 4.4). In the memorandum he summarized the evaluation of the geological risks mainly based on the final advice of the outstanding geology expert, Prof. Giovanni Lombardi from ETH Zürich. On the basis of his experience with other tunnelling projects and taking into account the test borings, he presented a list of six possible accidents ranging from 'small' accident (a cavern filled with water at low pressure, causing a delay of 3 weeks and costing CHF 100,000) up to a 'large' accident (a cavern filled with water under high pressure, delaying the project by 1 year with an additional cost of CHF 6–7 million). Since some geological faults, above all the Allondon fault, had been confirmed by the explorations of the Jura, he also pointed out that crossing a tricky fault could require an additional 16 months and CHF 17 million. With the tunnel under the Jura still at a depth of 600 m it would be extremely difficult to cope with a serious accident since access from the surface seemed impossible. The consequences of these considerations indicated a great risk regarding the time necessary to excavate the tunnel and even more so regarding the cost. In addition, the access to three of the eight experimental halls required access tunnels each with a length of 2–3 km, which would have made later operation rather cumbersome.

All these arguments led to a hazardous and gloomy prospect for the tunnelling and the whole project; hence, we decided to consider an alternative position for the ring although the formal proposal to Council was based on the actually foreseen position. This implied that we would have to convince Council to accept the new position even after the final approval. Fortunately that is exactly what we could achieve.

We decided to displace the main ring towards the east by several hundred metres, bringing it more out of the Jura and closer to the airport but still keeping it almost tangential to the SPS tunnel (Fig. 4.5). This new position offered several decisive advantages:

- The length of the tunnel under the Jura was reduced from 8 to 3 km.
- Three dangerous faults in the Jura could be avoided (although a major one still had to be crossed).
- The quality of the rock along the 3 km was better than that along the 8 km, thus reducing the risk.
- The maximum depth of the tunnel was reduced from 600 to 150 m.
- The largest expected water pressure would be only about 10 atm instead of 40 atm.
- In the case of serious difficulties, access from the surface would be possible in view of the reduced coverage of 150 m.
- The experimental halls could be accessed by vertical pits instead of special horizontal tunnels. This had particular significance in view of future projects in the tunnel. For LEP only four experimental halls were needed, whereas the new ring position made all eight access points directly available (an advantage which like all the others already mentioned turned out to be crucial for the LHC).

The new position took into account also a political aspect. The land needed for access shafts and surrounding installations had to be acquired by the two host states,

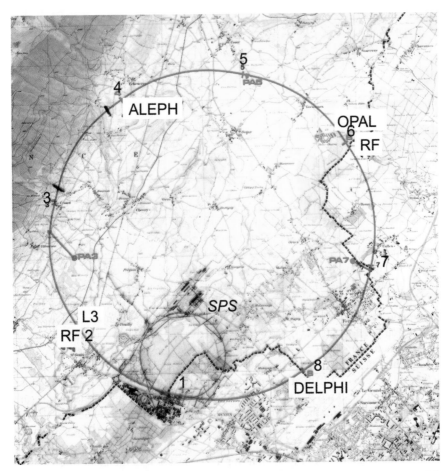

**Fig. 4.5** Final position of LEP. The *numbers* indicate the points with access shafts and straight sections. *RF* indicates the position of the accelerating cavities in straight sections 4.2 and 4.4. The *dotted line* shows the Swiss–French border. The positions of the experiments ALEPH, OPAL, DELPHI and L3 are indicated by their names

France and Switzerland. This could be handled in France by the relevant authorities. However, in Switzerland a popular vote is necessary for the acquisition of land by the governments (federal or cantonal), which might have taken a long time and the outcome was uncertain. Hence, the ring was placed in such a way that point 1, which is in Switzerland, was located in a region which had already been made available to CERN when the SPS was built and therefore no new decision needed to be taken. All the other access shafts were situated in France, with points 6 and 8 just a few metres outside Switzerland.

Most of the shafts were located in the molasse rock and could be excavated with conventional methods, e.g. by using explosives or cutting machines with mobile heads. However, before coming into contact with the molasse, several tens of metres of gravel had to be traversed.

**Fig. 4.6** Cross-section between points 4 and 8 showing the inclined position of LEP (1.4%, the vertical and horizontal scales are different). The position of the SPS is also indicated

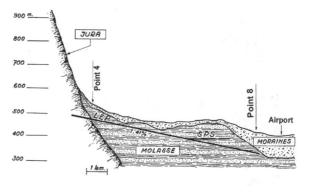

Of course, in real life it is not possible to gain only advantages; one always has to pay somewhere else. The geological surveys had shown that the roof of the molasse was descending towards the airport, where it was covered by almost 100 m of moraines. Whereas the molasse is a relatively solid rock and is excellent for tunnelling, the moraines are rather loose and very bad for excavation. To keep the tunnel in the molasse, it had to be rather low near the airport. On the other hand, we wanted it to be as close to the surface as possible at the foot of the Jura. These two conditions could be fulfilled only by putting the tunnel on an inclined plain with a slope of 1.4% (see Fig. 4.6). LEP was the first circular accelerator installed on a inclined plain. This seems to be a minor issue, however, because of the large size of the tunnel (diameter of about 10 km) even such a small slope results in a height difference of 140 m at two opposite points of the ring. One can even notice the slope when walking in the tunnel. This difference in height implies a pressure difference of 15 atm for the cooling water which has to be taken care of. The sloping tunnel also makes the installation and alignment of the machine components more difficult.

The final position also implied that for the connection of the main ring with the SPS a longer injection tunnel was necessary, which increased the cost somewhat.

However, these disadvantages were considered to be smaller inconveniences compared with the great advantages in reducing the excavation risks.

## 4.4 The Tunnelling Strategy

A tunnel of almost 27-km length cannot be excavated as a single section but has to be excavated in several sections. This was also necessary because of the completely different quality of the rock in the plain and under the Jura. The molasse has sufficient solidity so that immediately after excavation the walls and the roof of the tunnel are relatively stable. Therefore, a tunnelling machine with a rotating head boring the full tunnel section in one go can be employed (Fig. 4.7). Such tunnelling machines have the advantage that they can advance relatively fast, they do not damage the rock close to the tunnel itself and they do not shake the ground in the neighbourhood. A further advantage is that it is easy to take precautions against

**Fig. 4.7** Rotating head of the tunnelling machines. The front head carries rolls which break the rock

rocks coming loose from the roof of the molasse by use of prefabricated concrete vaults. Since the molasse is watertight and not soluble, no problems with water were expected. In addition, only a relatively small number of workers are needed in the tunnel, provided there are no breakdowns.

On the other hand, the fractured limestone in the Jura does not allow the use of such a tunnelling machine since in the case of an accident it could be squeezed in or even disappear into a cavern. A more conservative method had to be used: excavating with explosives. To avoid major incidents precautionary methods had to be taken, such as continuous pilot borings at the front of the excavation.

The total volume of rock which had to be extracted from the tunnel itself and also from the experimental underground halls and from auxiliary galleries amounted to about $1.4 \times 10^6 \, \text{m}^3$, about one third of the volume of the Cheops pyramid. Getting the spoil to the surface through only eight access shafts required a carefully planned schedule. First the rubble had to be transported from the place of the excavation through the narrow tunnel to a shaft and then lifted to the surface.

To speed up the excavations as much as possible, work at different places was planned in parallel. Originally two tunnelling machines were foreseen for the plain and a third one was added later. The tunnelling under the Jura was supposed to proceed independently. Of course, cost considerations were also a major issue.

The total work was divided for the final planning into the following sections (see Fig. 4.8) taking into account the results of the complicated tendering procedure.

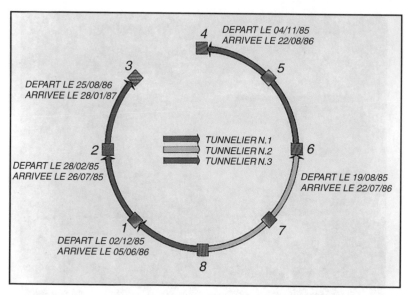

**Fig. 4.8** The tunnelling split up into different sections for three machines with rotating heads; the remaining part was excavated using explosives

In the plain:

1. Working zone A consisting of two sections using two tunnelling machines.

   (a) The first tunnelling machine starting the excavation from point 8 and going (clockwise) towards point 3 (9.4 km) with extraction of the rubble at point 8.
   (b) The second machine starting at point 8 and going (anticlockwise) towards point 6 (6.6 km) with extraction also at point 8.

2. Working zone B using a third tunnelling machine starting from point 6 and advancing towards point 4 (6.6 km). It was also foreseen that the excavation would continue beyond point 4 using small mobile-head excavation machines and explosives depending on the properties of the rock. A few hundred metres beyond point 4 the major geological Allondon fault was envisaged which should be approached carefully from both sides.

Under the Jura: Working zone C with the excavation starting from point 3 (using an access gallery starting from pit PA3) and advancing clockwise towards point 4, but only up to the geological fault, where it would meet the tunnelling from zone B (about 3 km away), crossing the fault being one of the most delicate operations to be executed. As mentioned above, the excavation under the Jura would have to be done by explosives.

One advantage of this planning was that all four tunnelling activities could be performed while *mounting* on the inclined slope of LEP. Working hill-up is always an advantage for underground work in particular as far as the break-in of water is concerned. It was expected that after a running-in period the tunnelling machines would advance on average by about 500 m per month, whereas tunnelling 'by hand' under the Jura would be much slower, about 150 m per month.

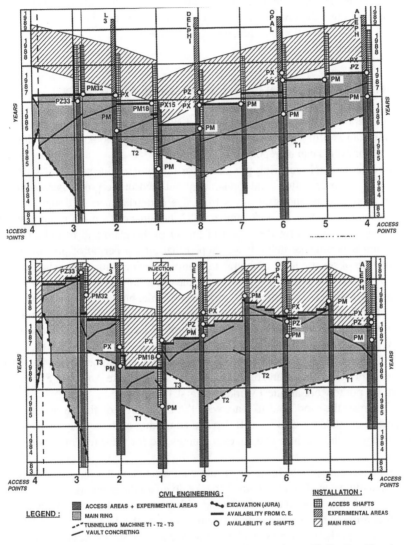

**Fig. 4.9** The excavation of the LEP tunnel, the shafts and the experimental halls. *Top*: The planned schedule. *Bottom*: The actual progress. The circumference of the ring is unrolled along the horizontal axis and the *numbers* correspond to the access points. The time is shown along the vertical axes

I shall not go into the details of the planning for the excavations of the experimental halls, the access shafts, surface buildings and some auxiliary galleries which had to be carefully synchronized with the excavation of the main tunnel.

All these and many more considerations were taken into account in setting up a time schedule including the preparations of the access shafts and experimental halls. It was shown that the civil engineering could be completed within 4 years, i.e. by the end of 1987 (see the dark horizontal bars in Fig. 4.9), assuming that the start would be late in 1983.

## 4.5  The Civil Engineering – Expectations and Reality

The full report of the civil engineering of LEP with all its ups and downs would sound like an exciting adventure story and would require a book by itself. Here only some major events can be reported.

The international tendering procedure for the civil engineering was started according to the preparations and the strategy outlined in the previous section. Seventeen groups representing 66 firms from five countries answered and after long and meticulous studies, on 4 November 1982 the Finance Committee approved the allocation of contracts for the 24 km of tunnel in the plain and including the underground experimental halls. It was hoped that after a period of preparation and the solution of several legal problems in France the excavation could start in spring 1983. A period of 4 years was foreseen for the execution of these excavations.

Some delays in the preparations were unavoidable, but finally 1 September 1983 was defined as the start date for the contracts which had been awarded to two international consortia: EUROLEP[2] for the work in the plain (zones A and B) and GLLC[3] for the excavation under the Jura (zone C) (see Sect. 4.4, Fig. 4.8).

On 13 September 1983 a ground-breaking ceremony took place in the presence of President François Mitterrand of France and President Pierre Aubert of Switzerland, whom we put to work as guest workers (Fig. 4.10).

Of course, the work had to start with the excavation of the 18 access shafts, varying in depth between 30 and 150 m and consisting of three types to bring down equipment for the machine and the experiments and for the safety of personnel:

- Eight machine pits to bring down the components of the machine, the widest with a diameter of 14 m at point 1 to take down all the magnets
- Five shafts for experiments, the largest with a diameter of 23 m at point 2, for the installation of the partly enormous components of the detectors
- Five access shafts for personnel, serving also as emergency exits

---

[2] Consisting of the five firms Impresa Astaldi (Italy), Entrecanales y Tavora (Spain), Fougerolle (France), Philipp Holtzmann (Germany) and Rothpletz Lienhart et Cie (Switzerland).

[3] Consisting of the six firms C. Baresel (Germany), Chantiers Modernes (France), CSC Impresa Costruzioni (Switzerland), Intrafor-Cofor (France), Locher (Switzerland) and Wayss et Freitag (Germany).

**Fig. 4.10** President François Mitterand of France and President Pierre Aubert of Switzerland laying the foundation stone of LEP on 13 September 1983

Most of the shafts were located in the molasse rock and could be excavated with conventional methods, e.g. by using explosives or cutting machines with mobile heads. However, before coming into contact with the molasse several tens of metres of gravel had to be traversed.

A special difficulty was met at point 8 near the airport where the molasse lies more than 100 m deep. Above it a water layer was found which serves as a kind of reservoir for the water supply of the population and no interference with it was tolerable. Various methods were tried to traverse it without water breaking into the pit. One method applied was to fill temporarily a circular trench, which later would become the pit wall, with a heavy liquid to keep the water out. In a second step the liquid was replaced by concrete. In some cases the only but expensive way was to freeze the ground. After the shaft had been excavated to a depth close to the water level (about 30–40 m deep), tubes were inserted into the ground around the whole shaft. A cooling liquid was pumped through them until the whole of the inner part was frozen to $-22°C$ (Fig. 4.11). This frozen block was then excavated and the water level was crossed under dry conditions. Three shafts had to be dug in this way.

The schedule for the tunnelling machines had to be changed relative to the original plans. Starting two machines at one access shaft (point 8) turned out to be too difficult even when this was staggered in time by several weeks. The availability of the access shafts for all kinds of operation had to be taken into account. In addition, efforts were necessary to compensate for some unavoidable delays. For example,

**Fig. 4.11** Upper part of a shaft with cryogenic equipment to freeze the ground

the staff of one contractor had organized two strikes asking for higher salaries, the first at the beginning of 1984, and the second in the summer of that year, which resulted in a total loss of 5 months.

The use of the tunnelling machines dictated by the circumstances mentioned was as follows. In February 1985 the first tunnelling machine was brought down at point 1 going towards point 2, where it arrived on 26 July 1985. Then it was transferred to point 6, from where it excavated the tunnel up to point 4. The second machine started at point 8, went in the direction of point 7 and continued to point 6. To compensate at least partially for the delays, a third tunnelling machine was employed contrary to the initial plans. Towards the end of 1985 this third tunnelling machine finally started work from point 8, excavating first the octant between points 8 and 1 and was then transferred to point 2, finishing the tunnelling in the plain at point 3 at the beginning of 1987. To avoid too much delay as a result of the unexpected events, the tunnelling schedule had to be changed completely and became much more complicated than the original planning, as can be seen by comparing the upper and lower parts of Fig. 4.9. In particular, looking at the lines marked by T1, T2 and T3 indicating the use of the tunnelling machines is revealing. The difference between hopes and reality!

A few details about the excavation might be of some interest. A tunnelling machine, called 'taupe' ('mole' in English) by French working people, is quite an impressive piece of equipment, resembling a small factory itself. The three machines for the excavation of the LEP tunnel were specially designed (Fig. 4.12) and adapted for the rock in the plain. The molasse is neither too hard nor too soft and is quite good for tunnelling. However, when it is in contact with air or water its hardness

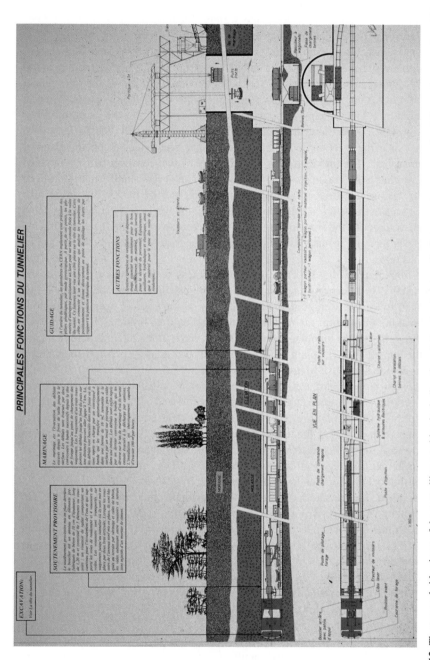

**Fig. 4.12** The top and side views of the tunnelling machine including the rotating head, automatic lining equipment, running belt to take away the rubble, and railway to bring prefabricated concrete elements and carry away the waste rock

changes quickly. Hence, the excavating machine must allow the immediate installation of concrete vaults to support the terrain. The head of the mole has a diameter of about 3.8 m, is about 10 m long and weighs 170 tons (see Fig. 4.7). It rotates about ten times per minute and carries 34 steel discs which break the rock. This front end is pushed forward by four cylindrical jacks with a total force of 460 tons. Behind the front head two telescopic shields are installed. The first one moves along with the cutting head, whereas the second is temporarily fixed against the tunnel walls to provide the necessary anchoring support for the hydraulic jacks. After an excavation of 1.2 m, which corresponds to the stroke length of the jacks, the fixed shield is moved ahead and the procedure is repeated. Directly behind these two shields a special gadget automatically puts in place prefabricated concrete parts of an annular vault. The broken rock is carried away by two conveyer belts and loaded on rail wagons and a train takes it to the bottom of an access shaft. After it has been lifted to the surface, it is stored to be taken away later by trucks. The trains also have wagons to transport installation material and the workers, since walking many kilometres to get to the working area was not acceptable. As the machines moved relentlessly forward, teams moved in to line and concrete permanently the excavated tunnel sections.

The tunnelling in the molasse went more or less without major problems. On 5 February 1986 a record was established by one machine boring 58.7 m in 1 day, more than twice the average figure. The effects of some minor accidents, such as rock breaking loose from the roof of the tunnel, could be limited. Of course, each breakthrough from the tunnel to one of the shafts was worth a celebration. In January 1987 the tunnelling in the plain was finished. This procedure is presently being used to excavate the tunnel for Metro Line C in Rome.

The excavation under the Jura had always been anticipated as the most troublesome section; it determined largely the time schedule and could have resulted in the whole project being a failure. To avoid major incidents, continuous pilot borings were made at the front of the tunnel before the main mass of the rock was broken loose by explosions. For each step of excavation more than 30 pilot borings were made systematically at the front of the tunnel (10–30 m ahead ) before the main mass of the rock was broken loose by explosions (Fig. 4.13).

In spite of all precautions a geological accident could not be avoided. On 3 September 1986 water from a geological fault with a pressure of 8.5 bar broke into the tunnel 15 m behind the front of the excavation with a flow rate of about 100 l/s. This demonstrated how tricky it was to bore into the rock. Pilot borings could not prevent a leak behind the actual place of excavation. Large parts of the tunnel were flooded (Fig. 4.14) and the excavation had to be suspended temporarily. In French such a leak is called a *renard* (a 'fox'), and many grim jokes about chasing the fox could be heard. However, we were all very worried about the continuation of the project.

Several methods to stop the water flow were tried. Injections of resins into the surrounding rock turned out to be difficult since the water was not static but flowing and it washed away the resin before it could harden. Injections into neighbouring rock up to a thickness of 3 m were made and when a second leak in connection with

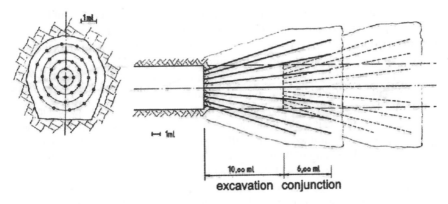

**Fig. 4.13** The arrangement of stepped pilot borings

**Fig. 4.14** The tunnel flooded by water under the Jura Mountains

the first leak was discovered additional injections became necessary. Eventually the water flow could be reduced and drained, and the roof and sidewalls of the tunnel were supported by iron and concrete vaults. Finally after 8 months the excavation could continue.

To avoid similar accidents it was decided to continue to inject resin at the front and around the tunnel systematically. The tunnel walls had to be lined immediately by spraying concrete on them to circumvent the detachment of rocks or the trickling of water. In addition, the tunnel walls were strengthened by a metallic vault to resist the water pressure. To make up, at least partly, for the delay a small tunnel leading to point 3, originally foreseen only for personnel, was equipped so that it could also be used for the underground work. Some minor infiltrations of water into the tunnel still occurred in the following years, necessitating some special measures, but they have not caused, until the present time, any major problems.

Let me mention a little dilemma which sounds funny but gave us some headaches. The water from the leak in the tunnel was not polluted but looked dirty, containing sand and fine clay. It drained into a small local river, the Allondon, which was used for fishing. The local anglers were very worried that the fish inventory might suffer and vehement protests followed. Before the French revolution hunting was a privilege of the nobles. Since then associations for fishing and hunting have been eager to preserve such rights for the 'people' and they became politically quite powerful in France. Hence, to avoid serious problems we had to take measures to clean the water by decantation before releasing it into the river.

Closing the ring of the tunnel at the fault of the Allondon near point 4 was a very delicate operation. The fault was approached from both sides. The excavation was continued from point 4 in the molasse and 80 m of tunnel was excavated, after which a mixture of molasse and clay was detected, implying that one was getting close to the known fault; hence, the work was stopped and one had to wait for the crew approaching the Allondon fault from the other side coming from point 3. The breakthrough finally happened on 8 February 1988 when the last explosive was detonated and removed a thin wall still separating the two ends of the tunnel. Emilio Picasso was waiting on one side of the wall, I was nervous on the other side and when the dust from the explosion had settled we cut a blue ribbon, which was the only obstacle left (Fig. 4.15). A little ceremony with 300 people in the underground hall of point 4, 150 m deep, followed and symbolized the formal end of the tunnelling.

**Fig. 4.15** The last obstacle to complete the tunnelling was to cut a blue ribbon (Emilio Picasso and Herwig Schopper)

## 4.6 Geodesy

Like for any civil engineering project, the surveying of the site was an important issue. However, the necessary precision for LEP presented a special challenge. For a road or a railway tunnel it may be acceptable if the track deviates somewhat from the ideal route. However, for LEP the geometry of the tunnel was strictly determined by the requirements of the storage ring (eight circular arcs with straight sections in-between). Errors in the alignment of the tunnel of more than a few centimetres were to be avoided since they implied additional excavations, with the resulting increase in cost.

To achieve such precision over the large underground distances of several kilometres required survey methods which at that time did not yet exist in Europe. One key element was the so-called Terrameter, an instrument developed in the USA for earthquake research. It was used for LEP for the first time in Europe and a precision of about $10^{-7}$ was achieved, i.e. the incredible accuracy of 0.1 mm for a distance of 1 km. To arrive at such a precision, the instrument uses two laser beams with different wavelengths, one red (0.6328 $\mu$m) and one blue (0.4416 $\mu$m), which permit the elimination of errors due to fluctuations of the temperature and pressure of the atmosphere.

Probably for the first time in Europe a satellite system, NAVSTAR, was used for geodesy in addition to the other methods. At that time such a global positioning system was still in its infancy for application in geodesy. It has the great advantage that no direct view is necessary between different survey points. It helped in an essential way to establish an extensive network of survey points in the whole region which was checked for its stability continuously and changes of less than 2 mm in the survey points were observed over several years. To understand the long-term behaviour of the system a model had to be developed taking into account the distortions of the terrestrial gravity by the large mass of the Jura.

To have a reliable network of survey points on the surface is one thing, but in the case of LEP it was essential to transfer the reference points from the surface down into the tunnel. This obviously was necessary to guide the tunnelling machines but later also to install the components of the machine. A simple plumb line would not do; because of the large diameter of LEP, even the curvature of the earth's surface had to be taken into account, the 'verticals' at opposite sides of the diameter not being parallel but converging towards the centre of the earth. In addition, the mass attraction of the Jura would cause deviations from the vertical.

Once reference points had been brought down into the tunnel, a further survey could be done with gyroscopes and lasers. Indeed the tunnelling machines were constantly guided by a laser beam. As a result, the real axis of the tunnel never deviated by more than 8 cm from the theoretical direction and the floor level was established with an error between 0 and $-1$ cm with respect to the theoretical slope. For a tunnel 27 km long, on a slope and with only eight access points 3.3 km apart this is an extraordinary achievement. The survey methods developed for LEP developed under the leadership of Jean Gervaise have certainly contributed to the development of modern survey methods in Europe.

## 4.7 The Arbitration

Everybody who has built a house knows that some conflicts with the contractor are practically unavoidable. This also happened for the civil engineering of LEP since the two consortia came up with additional requests going beyond the conditions established in the contracts. Most claims could be settled by negotiations led in a spirit of mutual understanding. However, the EUROLEP consortium was very obstinate. One might speculate about the reason. This consortium was led by the French firm Fougerolle, which had carried out many contracts for the French government or French state-owned companies. Apparently Fougerolle had become accustomed to making low bids during the tendering procedure and then during the execution of the contract asking for additional funds, which it normally obtained to avoid strikes. It tried to practise the same method at CERN, not being aware that at CERN we were bound to stick strictly to the contract conditions and had no freedom in granting additional compensation. When Fougerolle asked for additional funding, we recognized some of the requests as justified, others not. For example Fougerolle requested compensation for the losses due to the strike mentioned earlier. According to the rules such a request is justified if there was a general strike in the region, which, however, was not the case. Fougerolle argued that there was a strike at CERN at the same time. According to the employment conditions of CERN as an international intergovernmental organization, strikes are not allowed. Nevertheless, the Staff Association sometimes organizes a 'work stoppage' during Council sessions to impress Council with its arguments. Indeed one such 'work stoppage' occurred during the Fougerolle strike but had no relation to it. Of course, we could not accept the claims of Fougerolle in this case.

The negotiations with Fougerolle were extremely delicate for the following reason. We were prepared to grant Fougerolle a certain amount of compensation for some of the unexpected difficulties in executing the work but only for those which were well justified and in conformity with the contract conditions. On the other hand, we needed the agreement of the Finance Committee for any financial concessions. Yet, if we had asked the Finance Committee before the negotiations with Fougerolle, the latter would have learned what amount we were ready to concede and would immediately have asked for more. Consequently Picasso and I had to meet in full confidentiality with the president of Fougerolle and with poker faces we sometimes bargained for hours. Of course, we could not consent to a higher amount than what we thought was justified and what we could defend in front of the Finance Committee. Unfortunately in the end we had no success since Fougerolle refused a friendly settlement and brought the case to an international arbitration tribunal. After years of disputes[4] the outcome was that Fougerolle received compensation of about the same amount as we had offered in our confidential talks, but not more. Nevertheless, it should be underlined that all these disputes took place in an atmosphere of mutual respect and fairness.

---

[4] CERN Finance Committee 20 June 1990

## 4.8 What Else?

I should like to close this chapter by mentioning a few issues which are uncorrelated but sufficiently interesting to be reported:

1. Among the surface buildings were many conventional halls covering the pits and housing auxiliary installations such as ventilation and cooling. However, one building was very special. Since no preaccelerator for electrons and no facility to produce positrons existed at CERN, they had to be constructed and needed a building. To save funds it was decided to use old shielding blocks for the walls of this building which is 100 m long and has two stories. This building had to be located near the Proton Synchrotron (PS) accelerator, which would further accelerate the electrons and positrons to higher energies (see Chap. 6). Unfortunately no way was found to avoid interference of this building with one of the main roads at CERN and many people complained. Of course, we argued it would not remain in place forever, but this 'temporary' building is still in existence and is now used for accelerator research and development. The Empress of Austria, Maria Theresa, once remarked that the temporary solutions are the ones which last longest.

2. A human problem concerned the housing of the many hundred temporary workers coming from various countries for a limited time. When the civil engineering reached its maximum activity about 600 people for the civil engineering and partly overlapping with them 500 additional people for the installation of the machine had to be accommodated. Since there were not sufficient available

**Fig. 4.16** The people responsible for the civil engineering. From the *left* ?, Bruno Bianchi, Henry Laporte, Albéric Monjoie, Emilio Picasso, Mayor of the village crozet

hotel rooms or other forms of accommodation in the region, a number of special buildings had to be put up by the firms and the existing camping places had to be extended.

3.  In spite of the serious problems of geological and other natures mentioned in this chapter, the civil engineering was finished with a delay of only about 8 months in the plain and about 1 year under the Jura. That this was possible is largely due to the enthusiastic engagement of many people from the CERN staff but also from the firms involved. Above all, Henry Laporte should be mentioned as bearing the main responsibility for the civil engineering, supported by his deputy, Bruno Bianchi (Fig. 4.16). CERN and the contracting firms had done everything possible to provide a high level of safety for the workers by carrying out special studies and taking a number of measures. However, for a civil engineering project as large as LEP some accidents are unavoidable. Indeed two fatal accidents occurred, which is, thanks to the precautions taken, fewer than expected for a project of this size. We should remember those who gave their lives and show respect to their families for all their suffering.

# References

1.  CERN (1982) Etude d'impact du projet LEP sur l'environment, CERN, March 1982. CERN, Geneva

# Chapter 5
# The Environment – People and Nature

A large project such as LEP has considerable consequences for the environment. The impacts can be of different kinds – direct intrusion into the natural surroundings (water, climate, flora and fauna) and impairing human life (radiation, noise, anxieties). As with the assessment of all dangers, the problem is to find out whether they are real or based on wrong conceptions. At any rate we were determined to reduce any negative effects on the environment and on the population to a minimum.

## 5.1 Dialogue with the Population

To understand the worries of the neighbouring population and to convey to them the strategy of the project the most important action was to establish good communication. When I took over as director-general I learned to my surprise that my predecessors had followed a different policy. They thought it would be a hopeless attempt to try to explain to the general public the very complicated research and technological developments which CERN is undertaking. It turned out that because of the $N$ in the acronym of CERN[1] which stands for 'nuclear', many people thought CERN is a kind of advanced nuclear power station. Owing to lack of appropriate information and some other misunderstandings, a number of protests against the LEP project were raised.

To improve this situation we organized a campaign of 173 information meetings on French territory and several public conferences with explanatory talks and public discussions took place at the University of Geneva. Individual discussions were held with local authorities, e.g. mayors, delegates from various local or regional committees and other influential people. After these efforts a relatively small number of opponents remained, but they were particularly virulent. In some cases the reason

---

[1] CERN is the acronym for Conseil Européen pour la Recherche Nucléaire. To avoid misunderstandings it was considered changing the name of CERN. However, since the acronym CERN had become a quality mark, this idea was abandoned. Instead, in the letterhead of CERN 'European Laboratory for Particle Physics' is added.

H. Schopper, *LEP – The Lord of the Collider Rings at CERN 1980–2000*,
DOI 10.1007/978-3-540-89301-1_5, © Springer-Verlag Berlin Heidelberg 2009

may have been that opposing the LEP project brought immediate visibility to local politicians even if in reality they had no substantial arguments.

## 5.2 Radiation Safety – Hazards for the Population?

The impact of radiation on the environment is of great importance since it may present a major preoccupation or even danger for the population. It also has legal implications since legally LEP had to be considered as a nuclear installation[2] and required a certificate from the revelant authorities before it could start operation. In general, one can state that the radiation effects of a machine such as LEP, accelerating electrons (and its antiparticles, positrons), are much smaller than those of a proton accelerator or even of a nuclear power station. The reason is that the main interaction of electrons with the surroundings is only through the electromagnetic force, which is about 1,000 times weaker than the nuclear interaction of protons or neutrons. The main source of radiation for LEP was the emission of synchrotron radiation by the circulating electrons which is produced even without direct interaction between the accelerated particles and the surrounding matter.[3] As explained in Chap. 2, the intensity of the synchrotron radiation increases steeply with the energy of the circulating electrons. Another important parameter is the 'hardness' of the radiation (the photon energies), which determines its penetration power through matter. The so-called critical energy[4] of synchrotron radiation characterizes this 'hardness' and it increases with the third power of the electron energy. For the second stage of LEP operation (85-GeV beam energy) the critical energy was 400 kV and at the highest energy attainable with the LEP magnets (125-GeV beam energy, but never realized) it would have been about 1,300 kV. These are energies somewhat higher than in ordinary medical X-ray equipment and such radiation can be shielded against relatively easily.

One major troubling effect occurs if such radiation passes through air, where it produces toxic gases, mainly ozone and nitric oxides. This gave rise to violent opposition in public against LEP. A physicist who had previously worked at CERN and who knew the situation quite well claimed that on the basis of his calculations the production of toxic gases by LEP would be such that it would correspond to the exhaust gases of about one million cars and this would destroy most of the harvest in the region around LEP. Indeed the radiation power was quite impressive. For the

---

[2] According to French law every accelerator with a maximum energy above 1 GeV is classified as a nuclear installation (*installation nucléaire de base*).

[3] Most of the synchrotron radiation is produced in the bending magnets. In the straight sections there is no bending and hence no synchrotron radiation occurs. Some bending exists in the focussing/defocussing quadrupole magnets, but the synchrotron radiation produced there is negligible relative to that emitted by the bending magnets.

[4] The synchrotron radiation emitted from bending magnets has a very broad spectrum. The critical energy is defined in such a way that equal parts of the radiated power are emitted below and above the critical energy. Above the critical energy the spectrum drops exponentially.

two beams the emitted power was 2.5 MW at 50-GeV beam energy and 27 MW at 85-GeV beam energy. However, in his calculations he neglected to take into consideration some major facts. The vacuum chamber in which the particles circulated was surrounded by bending magnets along most of its length. The C-shaped yokes of these magnets enwrapping the beam tube (see Chap. 6, Fig. 6.2) consisted of iron and concrete and already represented considerable shielding. In addition, the aluminium vacuum chamber was surrounded by a special lead shielding which was 3-mm thick on the sides towards the magnet yoke and 8-mm thick on the side open to the air (Fig. 6.6). The combined result is that the radiation penetrating into the air was attenuated by large factors (about 1,000), leading to much smaller radiation powers (2.1 kW at 50 GeV and 360 kW at 85 GeV) [1]. Taking into account that ozone is not stable and that it takes the ventilation system some time to transport the gases to the surface, one can estimate that the pollution caused by LEP corresponded to the operation of a few additional motor cars! Much noise about nothing! Nevertheless, it took some effort and time to convince the public and the authorities.

Of course, other radiation sources had to be taken into account. Some of the circulating particles are continuously lost on the walls of the vacuum chamber, where they can produce radioactivity. In addition, the injector system produced radiation which had to be evaluated. These and many other effects were carefully studied, including a possible activation of the soil and water. The overall conclusion was that the total radiological impact of LEP on the environment and the population was insignificant and indeed so low that it was difficult to measure it, being masked by the fluctuations of the natural background. Nevertheless an intensive monitoring system for radioactivity and noxious gases was set up and this was permanently controlled by the competent Swiss and French authorities. During the whole existence of LEP no major radiation accident occurred.

## 5.3 Legal Problems

The legal situations concerning tunnelling under villages and private property are quite different in France and Switzerland. In Switzerland the rights of a private owner extend to a depth concerning possible private interests. These go to a depth of about 30–50 m, e.g. for the construction of underground garages or drilling wells. Since the LEP tunnel was deeper than this (about 150 m) we had no serious legal problems in Switzerland. Nevertheless there were some critical issues. For example, it was not completely clear whether a public referendum was necessary, which would have delayed the project considerably. Hence, I was very grateful to State Councillor Jaques Vernet, at that time responsible in the Canton of Geneva for public works, who called me on the last day of his mandate in June 1982 to his office to sign the formal approval for the construction of LEP. We had some doubts whether his successor, Christian Grobet, of a different political colour, would be as helpful, but to our satisfaction he also supported the project fully during the execution of the work. This is typical for Switzerland, where common interests prevail over pure party issues.

On the French side the legal situation was much more complicated and to obtain the authorization for the civil engineering took much longer. In France the owner of a property owns it all the way down to the centre of the earth (!) (however, excluding treasures of the ground such as oil or minerals, which belong to the state). The LEP tunnel passed under more than 2,000 private properties and an agreement with each of their owners was required. Since, in general, the tunnel was more than 100 m below the properties, we were happy we could get such agreements in most cases by friendly negotiations. In a few cases where the properties were close to an access shaft, the owners might have felt that they were entitled to some kind of compensation for nuisances they were expecting. Some owners could be satisfied by the putting up of earth dams which protected them from noise and ugly sights or by the provision of better access roads to their properties.

In addition, the owners had the right to compensation for conceding part of their territorial right allowing the tunnelling under their property. Since France as one of the host states of CERN had agreed to provide the land free of charge, the French state had to pay such compensation. However, the amounts offered were rather small and hence some owners wanted to start legal procedures against CERN to obtain better compensation. To sort out all these legal problems almost 2 years was needed between the approval of LEP by the CERN Council in October 1981 and the ground-breaking ceremony in September 1983.

According to the site agreements with the two host states[5] and following international practice, an international governmental organization cannot simply be sued like a national organization. If somebody wants to make a claim against CERN he has to start a complicated procedure going through the national legal authorities, which have to follow the rules of international public law. Another possibility is to settle the conflict by an amicable arrangement, which had been used in practically all cases in the past. Some people concerned, however, were not satisfied with this situation and requested that CERN renounce its rights corresponding to its international character. I refused to give in to such requests since if CERN had given up this privilege in one case it would have been lost forever. Nevertheless, if only one proprietor had started a legal procedure, if not against CERN but against the French government, this could have delayed the project enormously.

Legal judgements usually take a long time and waiting for a decision, a court might even have stopped the continuation of the tunnelling. To avoid this we had to apply for a *declaration d'utilité publique* ('declaration of public interest') to be issued by the highest French court, the Conseil d'État. However, to obtain such a 'green light' a careful study of the whole environmental impact of LEP had to be made, the *étude d'impact* which will be described in Sect. 5.4. Before such a document could be submitted to the French authorities, public hearings had to take place, which happened from 13 September to 5 November 1982. All the persons concerned could express their objections. The regional commission dealing with the operation of real estate had to give its opinion before the issue could be finally presented to

---

[5] Agreement with the Federal Swiss Council dated 11 June 1955 and with France dated 13 September 1965, revised 16 June 1972

the Conseil d'État. The legal advisor to the director-general, Jean-Marie Dufour, had obviously a lot of work to do and he did it in an extraordinary manner. We were also very thankful for the support we received from various local authorities such as the General Council of the Département of Ain, many mayors of the communities in the region, several prefects and sous-prefects who followed each other successively during this period. Last but not least the continuous support of Senator Roland Ruet, well respected in the region and in Paris, was crucial. Finally on 20 May 1983, on the basis of the court decision, the French prime minister issued the requested declaration of public interest and the ground-breaking ceremony took place on 13 September 1983.

## 5.4   The Environmental Study – *Étude d'Impact*

One of the crucial elements which had to be provided before the construction of LEP could start was an environmental study – or *étude d'impact* according to French legislation. Indeed it was also in the interest of CERN to have such a study which included a comparison of the environment before the project started with that to be expected after the execution of the project. Such a study seemed necessary to establish good relations with the population and the authorities in the region and to protect the organization against later legal claims. The worries went in two directions: disturbances during the construction (e.g. road traffic of heavy trucks, pollution of rivers) and troubles during operation (e.g. draining underground water reservoirs leading to a lack of drinking water, noise, unpleasant views).

Robert Lévy-Mandel, a very competent French engineer and administrator, having spent many years at CERN was asked to establish the environmental study with the help of CERN and outside specialized services. This was an enormous task, resulting in a document [2] of about 170 pages and covering much-diversified topics such geography, geology, hydrology, climate, flora and fauna, living conditions and cultural activities of the local population, traffic and water supply. In each case the influence of the LEP project had to be considered and foreseen measures to keep disturbances low had to be explained. To stay in close contact with the local authorities and politicians, Lévy-Mandel setup a structure of conciliation which served to avoid problems due to possible misunderstandings or lack of communication.

Of course, it is not possible to deal here with all the areas mentioned. However, I would like to report on some of the major problems which came up.

The local region between the Jura and the French–Swiss border is called the Pays de Gex, with the little town of Gex as Sous-Prefecture. It is lovely countryside, mainly of agricultural nature. A general preoccupation was that LEP would change the characteristics of the region into an industrialized zone. The slogan was going round that "in the past CERN was in the Pays de Gex, now the Pays de Gex would be inside LEP". This worry, geographically almost true, was greatly exaggerated. Indeed when one takes off or lands at Geneva airport it is very difficult to detect the few surface buildings which had to be erected for LEP and this is even true at present for the additional LHC buildings.

One anxiety which was difficult to eradicate concerned the supply of drinking water. It had happened in the past that during very dry summers some of the villages had a lack of water and some people were afraid that the underground reservoirs might be depleted because of LEP. The study of the hydrological features had shown that this was practically impossible, but sometimes perceptions cannot be changed by facts. To avoid problems which could have delayed the whole project, we had to agree to connect a few villages to better water sources. This even had amusing effects – some villages farther away from the final LEP position requested that we move the tunnel closer to them so that also they could ask for a better water supply system.

Another issue which led to complicated discussions concerned the storage of the rock to be removed from the underground: where could the $1 \times 10^6$ m$^3$ be stored and how would be it transported? A big truck can take about $10$ m$^3$, which implies that about 100,000 truckloads had to be transported within about 2 years. What made it worse was the fact that the traffic could not be distributed uniformly over time but at some peak times between 100 and 150 trucks had to depart each day from some access shafts. In a rural area with small roads through little villages this resulted in the population becoming greatly annoyed. To keep the disturbance to a minimum, the first measure was to keep the distances between the shafts and the deposits as small as possible. In cooperation with the local authorities, places for the deposits could be identified which indeed were close to the access shafts. In addition, it turned out that the waste could be used in positive ways. It served partly to fill up abandoned quarries and partly to turn swamps into green land. To put all these plans into reality it was necessary:

- To build some private CERN roads linking the access shafts to the public road network
- To strengthen and widen some of the communal and departmental roads
- To build some roads together with the French authorities to bypass some settlements

In total about 39 km of roads was concerned and the financing of the improvements had to be negotiated with the local and national authorities. The total investment amounted to FFR 75 million, which was distributed between the French central government, the local Département and CERN according to the ratios 50, 35 and 15%, a solution acceptable to CERN. In that way the Pays de Gex even benefited from LEP in the long run by getting better roads.

## 5.5  Energy Consumption

Of course, CERN had always been aware of the necessity to keep energy consumption to a minimum, be it for environmental reasons or for financial ones. For LEP the equipment dominating the need for energy was the accelerating system, which not only had to accelerate the particles, but which had to continuously compensate for the large losses due to synchrotron radiation. Ingenious ideas were applied to keep this consumption as low as possible (see Chap. 6). When all the other components were designed, the energy efficiency was also taken into account.

As a result, the maximum total electric power for LEP amounted to about 70 MW in the first phase, somewhat less than the proton accelerators at CERN had required, and it went up to about twice that value for the final operation mode of LEP. There was no problem to get this energy transported to the site since CERN had already been connected to a French high-power 400-kV line getting its power from a nearby hydroelectric station.

To keep the load on the public net as low as possible, the following agreement was concluded with the French electricity company EDF. On cold winter days the French power net has to cope with very high peak demands since electric power is used in France largely for heating purposes. A large fraction of electric power in France is provided by nuclear power stations, which are good at providing a constant base load but cannot easily provide peak power. Therefore, in wintertime France is sometimes confronted with power shortages, the 'critical days'. To help reduce the consequences of such incidents, we agreed to shut down the operation of LEP at short notice for a total number of days during one season (about 10 days). This presented, on the other hand, a serious problem for the CERN users who came from their home universities to perform experiments at LEP and had to accept that during their stay at CERN the machine was stopped without previous warning. However, we could reduce somewhat the negative consequences of this agreement. Most of the facilities at CERN and above all the big machines need a yearly maintenance shutdown to guarantee their good performance. By scheduling this shutdown during the winter months, we could ensure that many of the critical days would fall in the shutdown period and not trouble the scheduled operation of LEP.

In summary, one can state that the peak power needed and the average power consumption of LEP per year created only a minor impact on the environment thanks to the measures taken to minimize this consumption, including the reduction of other CERN programmes and owing to general energy saving schemes. The total yearly consumption of electricity by CERN before the construction of LEP sometimes surpassed 650 GWh/year. During the years when LEP was in full operation in the high-energy range it went up eventually to about 800 GWh/year. This is certainly a considerable amount of energy, equivalent to the consumption of a little town. Compared with the total consumption of Geneva (without CERN), which is about 2,700 GWh/year, it would seem tolerable in view of an international project serving many thousands of users from all over the world.

## 5.6 Additional Measures

Many other measures were taken to minimize the problems for the population around CERN. In agreement with and at the request of the people in some particular areas, dams were raised to reduce noise or cover unpleasant views, trees and bushes were planted and buildings and cooling towers were designed to be as low as possible or were erected in local depressions. A special action was undertaken by creating a nursery of young trees. About 1,000 saplings were raised for about 5 years and then transplanted to the access areas of the experiments to harmonize the surface buildings with the landscape.

It would take too much room here to describe all the additional actions which were taken to keep the relations with the neighbours as friendly as possible. Good contacts with the local authorities were essential and the help of G. Mazenot, Prefect of the Ain, L. Ducret, Mayor of St.-Genis-Pouilly, J. Raphoz, Mayor of Prévessin-Moens, and Senator Roland Ruet were much appreciated. When at the end of the civil engineering I met the mayors of the neighbouring villages at lunch (in France it is always better to meet people at meals!), they all expressed their satisfaction with the way we had handled their worries and they admitted that the economic advantages which LEP brought to the Pays de Gex overcompensated for the negative effects.

Of course, there are always some people who use dishonest arguments just to get some benefits for themselves. A local politician tried to organize an anti-LEP movement to get more publicity and influence in local politics. He was a partisan of the opposition, claiming that LEP would destroy the agricultural environment of the region. However, after LEP had been in operation for some time he wanted to sell some property and advertised it pointing out its particular advantage of being close to LEP and CERN! Another example was a bright student who had worked for some time at CERN and who understood perfectly what the implications of LEP and the general conditions at CERN were. Apart from environmental issues, he criticized CERN for being involved in military research and stated that LEP might intensify such activities. However, he knew quite well that CERN according to its Convention is not involved in any classified work, that nothing is secret and that indeed the many users from many countries provide the best guarantee that this principle is strictly respected.

LEP, like previous projects at CERN, has proven to the population on both sides of the Swiss–French border that CERN makes great efforts to interfere as little as possible with the environment and no major problems arose during the 12 years of LEP operation. On the other hand, there is no doubt that the region has benefited economically from LEP and CERN in many ways.

# References

1. Goebel K (ed) (1981) The radiological impact of the LEP project on the environment, CERN 81-08, 20 July 1981. CERN, Geneva
2. CERN (1982) Etude d'impact du projet LEP sur l'environment, CERN, March 1982. CERN, Geneva

# Chapter 6
# LEP – The Technical Challenge

LEP was the culmination of a development which continued over several decades, starting with very small electron–positron colliders, such as ADA in Italy fitting into a normal room, and ending with LEP, the largest research instrument ever built. Therefore, some people think that technically LEP was just a blown-up version of earlier machines such as PETRA at DESY or TRISTAN at KEK in Japan and hence that no major technical innovations were necessary. This is a completely wrong perception. Because of its huge size and the resulting cost, many technical innovations were necessary to realize the project, sometimes implying considerable technical and financial risks. Only the most spectacular developments can be mentioned here.

Following the dramatic evolution of the project and the choice of the size and position of the LEP tunnel described in Chaps. 2 and 3, the detailed design of the facility could be finished. The guiding policy issues were:

1. Use the existing accelerators as preinjectors
2. Realization of the project in three major phases

    (a) LEP 1 with a maximum beam energy of 55 GeV to produce the Z particles in large quantities ('Z factory') using copper rf accelerating cavities.
    (b) LEP 2 with a maximum beam energy of 100 GeV and possibly beyond to produce W particles and other new particles by using superconducting rf accelerating cavities.
    (c) LEP 3 to install later a superconducting proton ring in the tunnel either for electron–proton collisions (Super-HERA following the HERA collider at DESY) or for hadron (proton–proton or proton–antiproton) collisions. The idea to have both an electron ring and a proton ring in the tunnel was later given up; the electron ring was dismantled and LEP 3 became the LHC.

3. Finding technical solutions which would allow the realization of the project within the strict budgetary and work force limitations imposed by Council.

H. Schopper, *LEP – The Lord of the Collider Rings at CERN 1980–2000*,
DOI 10.1007/978-3-540-89301-1_6, © Springer-Verlag Berlin Heidelberg 2009

To supervise the project a LEP Management Board was created. It was chaired by the project leader, Emilio Picasso, and its members were the leaders of the various subsections of the project (Roy Billinge, Franco Bonaudi, Henry Laporte, Bus de Raad, Hans-Peter Reinhard, Lorenzo Resegotti and Wolfgang Schnell), the LEP division leader, Günter Plass, and the director responsible for the other accelerators, Giorgio Brianti. Picasso managed to maintain through the whole period of the construction of LEP an extremely cooperative spirit in the LEP Management Board. Of course, sometimes clashes of interest were unavoidable, but they were always solved in a fair and constructive way. I insisted on attending the meetings of the LEP Management Board, not to interfere with the technical decisions to be taken, but mainly to be well informed about the requirements of the project. When later in the CERN Directorate it was necessary to decide on the allocation of financial and personnel resources I had a much better feeling of what LEP really needed.

To benefit from the best expertise in the world an international LEP Machine Advisory Committee was set up which was chaired by Gustav-Adolf Voss from DESY, one of the most experienced accelerator experts. This committee accompanied the construction of LEP by providing extremely useful and critical advice.

## 6.1 How Does a Collider Work?

Before going into the specific LEP design, it may be useful to describe here the principle of the functioning and the main components of an electron–positron collider, which are shown in Fig. 6.1. Packets of electrons and positrons (in the case of LEP four packets for each kind of particle) are accelerated in a preaccelerator chain and injected into an annular vacuum chamber which has to be highly evacuated so collisions with air molecules are avoided. In the chamber electrons and positrons run around in opposite directions and meet at eight collision points. The particles are kept on a circular track by C-shaped dipole magnets, simply called 'bending magnets', covering as much of the circumference as possible . As explained in Chap. 2, the energy of the electrons is not limited by the strength of the bending magnetic field but by the loss due to synchrotron radiation. To keep the latter low, any abrupt curvature must be avoided and the magnets should guide the particles around as smoothly as possible.

Energy has to be supplied to the circulating particles, first to accelerate them to the required energy and then to compensate for the continuous synchrotron radiation losses. For this purpose, rf cavities are installed in some of the straight sections of the ring and these produce an electric field to push the particles at the right moment when they traverse the cavities. Synchronism has to be established between the circulating particles and the alternating high-frequency accelerating field (hence the name 'synchrotron' for this kind of accelerator) to achieve an energy gain. In Fig. 6.1 copper accelerating cavities as used in LEP 1 are indicated. They were replaced in LEP 2 by low-loss superconducting cavities.

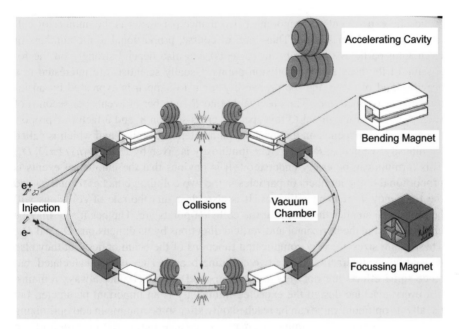

**Fig. 6.1** An electron–positron collider showing its main components

In order not to lose particles by their hitting the vacuum chamber wall, but rather to let them circulate for hours, special focussing magnets have to be installed which keep the particles close to the centre of the vacuum chamber. The motion of the particles travelling through these focusing elements resembles in some aspects that of light rays being focussed by the lenses of a camera and in analogy one speaks of the 'beam optics'. Very similarly to the use of lenses in a camera objective, different kinds of lenses are used to correct various kinds of errors (chromatic, aspherical) and the design of the beam optics has become a special art requiring complex computer programs. To focus electrically charged particles one uses quadrupole magnets, which consist, as the name indicates, of four magnetic poles and poles of the same polarity face each other (shown in Fig. 6.1). To correct for focussing errors (e.g. owing to slightly different energies of the particles corresponding to chromatic aberration in the optics) one needs, in addition, sextupole magnets. A set of the three magnets (bending, quadrupole and sextupole) forms a unit cell. The LEP magnet system had eightfold symmetry around the circumference, which means that there were eight identical octants each consisting of 488 unit cells. In addition, a number of special magnetic and electric steering elements are needed to control the beams. If everything works properly, the particles perform oscillations (the so-called betatron oscillations) about the centre of the vacuum chamber with amplitudes of a few centimetres, such that the particles do not touch the walls of the vacuum chambers. Many other elements for observing and steering the beams are necessary, but they cannot be discussed here.

For the experiments an important performance parameter is the number of collisions produced per second. These are, of course, proportional to the numbers of circulating particles (i.e. the beam currents), but also depend strongly on the focussing of the beams at the collision points. Usually scientists are interested in a specific kind of event and the probability for it to happen is expressed by an interaction cross-section $\sigma$. The observable rate $R$ (number of events per second) of such events is proportional to this specific cross-section $\sigma$ and a factor of proportionality which depends on the machine parameters mentioned and which is called 'luminosity' $\mathcal{L}$; hence, $R = \mathcal{L}\sigma$. The luminosity is given by $\mathcal{L} = f n_1 n_2 / 4\pi D_x D_y$. This formula can be easily understood. It is obvious that the number of events is proportional to the numbers of particles in the two colliding bunches $n_1$ and $n_2$ and the frequency $f$ of their collisions. It is also clear that the rate of collisions will be higher the smaller the beams at the collision points are. The beam size may be characterized in the horizontal and vertical directions by its dimensions $D_x$ and $D_y$. These beam sizes are very complicated functions of the beam optics, but they also depend on the interaction of the two colliding beams. They can be calculated, but since some effects are not very well understood, an uncertainty always remains. For every machine design the expected luminosity is an important parameter, but usually its optimum value can be reached only after some running in and optimizing of the beam optics. The luminosity may also change during the operation of the machine and therefore it has to be continuously measured and monitored. The rate $R$ gives the number of events per second. In an experiment, of course, the total number of events occurring during the whole period of data-taking matters and is proportional to the number of operating hours of the machine, of course. The potential for new discoveries with a collider depends on the maximum achievable energy and on the number of observed events, which is determined by the luminosity. Hence, after the maximum energy, the luminosity is the most important parameter of a collider.[1]

Following the general guidelines of the Pink Book (see Chap. 2) the final parameters of the facility were laid down in the *LEP Design Report* consisting of three volumes: volume I [1], *The LEP Injector Chain*, volume II [2], *The LEP Main Ring*, and volume III, *LEP2* [3].

Some of the design parameters of LEP 1 are given in Table 6.1 (see [1–3]). The parameters for LEP 2 are shown in parentheses and will be discussed in Chap. 14. The many digits for some of the numbers indicate the enormous precision which had to be achieved to ensure the good performance of the machine. It can also be seen that the beams at the collision points are not round but flat bands.

---

[1] Most of the explanations given here for an electron machine are also valid for a proton collider except that the beam dynamics is different owing to the absence of synchrotron radiation (no damping of oscillations).

**Table 6.1** Main design parameters of LEP 1 (in *parentheses* those of LEP 2)

| | |
|---|---|
| Circumference | 26,658.883 m |
| Bending radius in dipoles | 4,242.893 m |
| Number of bunches per beam | 4 |
| Number of interaction points | 4 |
| Ratio of Horizontal/vertical beam width | 25 |
| Frequency of rf cavities | 352.20904 MHz |
| Power for rf cavities | 16 MW |
| Revolution time of particles | 88.92446 microsec |
| Active length of accelerating cavities | 272.4 m (817 m) |
| Total accelerating voltage/circumference | 400 MV (4,100 MV) |
| Injection energy | 20 GeV |
| Maximum beam energy | 60 GeV (105 GeV) |
| Peak luminosity | $1.6 \times 10^{31} \mathrm{s}^{-1}$ |

## 6.2 The 'Concrete' Magnets

As explained in Chap. 2, the magnetic field that keeps the electrons on a circular path (the main magnet ring) need not be high in an electron storage ring.[2] The important issue is that it is distributed uniformly around the ring. Any abrupt bending leads to synchrotron radiation losses. For LEP 1 only a field strength of 0.135 T was required for an energy of 65 GeV. Even for LEP 2, with about twice the energy and a corresponding magnetic field twice as high, these are relatively small magnetic fields for an iron magnet (about 10 times weaker than in accelerators for protons). To cover 20 km of the LEP circumference with costly iron magnets to produce a low field seemed an enormous waste of funds, although for machines such as PETRA at DESY or TRISTAN at KEK with circumferences of several kilometres iron magnets were still a reasonable solution.

For LEP an ingenious new technology was invented for the main ring magnets, the so-called concrete magnets. They have the conventional C-shaped yokes (Fig. 6.2) and each magnet is 5.75 m long. However, instead of using compact iron, the yokes are composed of a stack of low-carbon steel laminations, 1.5 mm thick, separated by 4-mm gaps.[3] Such an amount of iron is sufficient to produce the necessary magnetic field; however, because of the extreme accuracy needed for the field quality the magnet structure needs enormous mechanical stability. This was achieved by filling the gaps between the iron laminations with

---

[2] In a proton machine the maximum energy is determined by the strength of the magnets, whereas in an electron machine the rf power to compensate for the synchrotron radiation losses limits the energy.

[3] This geometry gives an iron filling factor of 0.27 with a drop of 5% of the ampere-turns at the maximum field, which turns out to be an economic optimum.

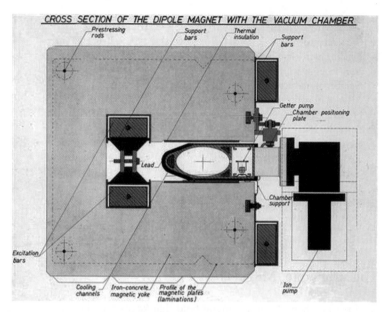

**Fig. 6.2** Cross-section of a bending magnet, showing the C-shaped yoke, the vacuum chamber and the excitation bars

cement mortar. Each magnet was terminated by two end plates which were held together by four prestressed rods. With such a design, a magnet behaves like a pre-stressed concrete beam, whose general properties are more or less known from civil engineering.

However, the accuracy necessary for LEP was higher than what is required in civil engineering. Extensive research was carried out to find the best mortar composition, to avoid shirking on drying out and to ensure low risk of corrosion of the embedded steel. A special trick was used to reduce long-term deformations due to creep. The forces on the prestressing rods were adjusted in such a manner as to produce a varying compressive force, being twice as high at the back of the yokes as at the front. This asymmetry counteracted the horizontal bending (about 2 mm after 3 years) resulting from the different shrinkage between the back and the front of the yokes. Several years of development carried out in cooperation with civil engineering firms was necessary to learn how to produce these magnets. In the end some new knowledge was also obtained concerning the construction of concrete beams for civil engineering. It was not known at the beginning what energies would be needed for LEP 2. To be on the safe side the bending magnets were designed and constructed for a maximum energy of 125 GeV. As will be reported in Chap. 14 (see Fig. 6.10), such energies were unfortunately not attained because of the limitations of the rf power.

All these complications represented, of course, a considerable risk. It would have been a disaster if after a few years LEP had not worked anymore because of the

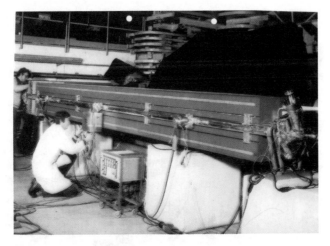

**Fig. 6.3** A 'concrete' magnet during testing

deterioration of the bending magnets. No other accelerator had used such 'concrete' magnets before, but we hardly had a choice; conventional iron magnets could not have been afforded within the available budget. Indeed the cost of the concrete magnets was about only half that of conventional magnets. Fortunately it turned out that during the 12 years of operation of LEP these concrete magnets did not present any problems.

A finished magnet painted red looked like a normal iron magnet (Fig. 6.3) and only on close inspection could one see the iron–concrete laminations. In total 3,392 concrete magnets, each 5.75 m long, had to be installed in the tunnel. Their mass production and testing before the installation required a considerable organizational effort. Since the magnets were produced before the tunnel was available for their installation, they had to be stored and the old Intersecting Storage Rings (ISR) tunnel was used for this purpose. Figure 6.4 gives a vivid impression of the kind of mass storage one had to deal with. Transporting the magnets from the storage site to the access shafts and finally into the tunnel presented a remarkable problem of organization, which was solved in a spectacular way. The final places of the magnets in the ring were sometimes several kilometres from the nearest access shafts and a special rail system suspended from the tunnel ceiling was used, as will be described in Sect. 6.5.

The iron yokes have to be excited by an electric current to produce a magnetic field. This field has to increase while electrons are accelerated from the injection energy of 22 GeV (see Sect. 6.1) and to keep them on the same circular path during the energy increase, the magnetic field has to grow in synchrony.[4] When the electrons have reached the required energy, the field must be kept constant to store

---

[4] This is one reason why such accelerators are called 'synchrotrons'. The necessary synchronism between the circulation frequencies of the particles and the rf accelerating field was mentioned in Sect. 6.1

**Fig. 6.4** The storage of the bending magnets in the old Intersecting Storage Rings (ISR) tunnel

them for as long as necessary. In a normal accelerator each magnet has its own coil (usually with several turns) to excite the magnetic field. For the low magnetic field of the LEP bending magnets a single-turn coil would have been sufficient. Indeed, in principle, a single common conductor passing through all the bending magnets around the ring could have been used, resulting in considerable savings. However, with such an option LEP would have produced a geographically extended magnetic field filling the whole interior of the LEP ring and reaching into the atmosphere above LEP. Although its strength would have been quite low, it might have changed the colours displayed on TV sets in the Pays de Gex and it could have interfered with navigational systems of the nearby airport. Hence, a second conductor, a return bar, had to be added, concentrating the magnetic field to the gaps of the C-shaped yokes. In practice the current for the excitation was distributed among two excitations bars on the inside of the magnet yokes and two return bars on the outside (see Figs. 6.2, 6.5), but still forming large loops around all the magnets of one octant of the ring. The conductor bars with dimensions of 90 mm × 44 mm were made of aluminium and had to carry an electric current at the maximum energy of LEP 1 of about 2,700 A and 4,500 A for LEP 2. The power dissipation in the latter case was about 16 MW. Hence, the conductors had to be cooled and for that purpose each aluminium bar had a central hole through which cooling water was pumped.

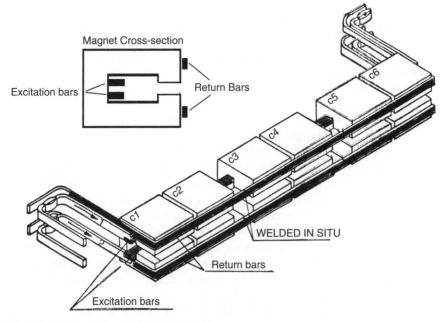

**Fig. 6.5** The electrical excitation of the bending magnets

This unconventional scheme of exciting the main bending magnets by common aluminium bars instead of individual coils for each magnet reduced the cost appreciably. Of course, not using the return bars would have resulted in additional savings since their cost simply increased with the circumference (many tons of aluminium would not have been necessary). However, to maintain good relations with the neighbouring population, the more expensive solution was adopted.

In addition to the bending magnets, other magnetic elements were necessary. To focus the particles and keep losses to the vacuum chamber walls low, 816 magnetic quadrupoles and 504 sextupoles had to be installed. Special magnets were used to squeeze the beams at the collision points; other elements served to steer the beams or to inject them into the main ring. These elements could be constructed using existing techniques and therefore will not be discussed here any further. However, it should be stressed again that because of the size of LEP and the great number of elements, the control of the beams during acceleration and during storage became quite intriguing.

## 6.3 The Vacuum System

Electrons and positrons have to circulate in a chamber with an extremely low air pressure since otherwise they would collide with air molecules instead of colliding with each other. Hence, a vacuum chamber had to be provided along the whole circumference of 27 km. A further complication comes from the fact that the circulating particles emit intense synchrotron radiation, which hits the chamber walls and frees molecules absorbed there. This effect is certainly greatest at the beginning of operation when many air molecules are still adsorbed at the walls, but some outgassing continues during the whole operation. Hence, a very powerful pumping system is required which can not only extract the air from the vacuum chamber after closing it, but is also capable of constantly removing the desorbed gases.

This challenge had not been met before for a vacuum chamber 27 km long; therefore, in this case new imaginative solutions had to be found. The individual sections of the vacuum chamber were fabricated by extrusion from aluminium and had a complicated cross-section (Fig. 6.6) with wall thicknesses between 3 and 8 mm. The elliptical portion for the beams had dimensions of 13 cm wide and 7 cm high. This part had to be cooled to carry away the power deposited by the synchrotron radiation (several megawatts) and three cooling channels were integrated into the aluminium profile. To get rid of the desorbed gases during operation along the whole path of the circulating particles it was necessary to use distributed pumps along the entire beam chamber and as close as possible to the beam. For this purpose a rectangular pump section was attached to the elliptical beam chamber and the two parts were connected by many holes in the chamber wall.

The distributed pumping was achieved for the first time in an accelerator by so-called getter pumps. The principle of getter pumps uses the properties of some metals, e.g. zirconium, to absorb eagerly most of the gases in the atmosphere. To activate the pumping, the metal has to be heated to about 700°C to clean it before starting normal operation and after some time when it has become saturated with gases it has to be reconditioned by heating it to about 400°C. Such getter pumps

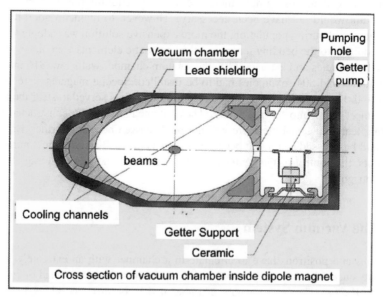

**Fig. 6.6** Cross-section of the aluminium vacuum chamber

were used in most conventional TV tubes. For LEP a getter band about 20 km long was needed for the gettering and a low-cost solution had to be found. In cooperation with industry, a new alloy consisting of 84% zirconium and 16% aluminium was developed and became known under the name 'non-evaporable getter' and is now used in many accelerators and other applications.

To avoid early saturation of the getter band, a primary vacuum had to be produced by conventional pumps in discrete locations. Almost 3,000 pumps of different kinds were needed for that purpose. With this powerful vacuum system a static pressure of about $10^{-10}$ Torr could be achieved, a vacuum comparable to that in cosmic space. Such extremely low pressures had been achieved before in many laboratories but only in small vessels, never in a container with the dimensions of the LEP vacuum chamber. Even in operation, when the outgassing due to the synchrotron radiation was taking place, a low pressure of $3 \times 10^{-10}$ Torr could be maintained. This is very crucial since the 'lifetime' of the circulating particles is proportional to the amount of residual gas. With the relatively simple but innovative solution chosen for LEP, electrons and positrons had to be refilled only after many hours.

The total length of the vacuum chamber had, of course, to be fabricated in shorter pieces, with a length of 12 m each. The production of the chambers, respecting very severe tolerances, was pushed to the limit of existing technological possibilities and required many trials and tests. A particular problem was the cladding of the aluminium chambers with lead, which was necessary to absorb the synchrotron radiation. Since an essential fraction of the power was absorbed in the lead shielding, this had to be cooled. For that purpose the bonding between the lead and the cooled aluminium of the vacuum chamber had to be very tight and effective over most of the area where the two metals touched. To achieve good bonding a thin layer of nickel was put between the aluminium and the lead. Since nickel is a magnetic material, this layer caused some unexpected changes of the magnetic field guiding the beams and it took some time to recognize the origin of the 'miraculous' behaviour of the beams. Summarizing, one can say that the construction of the vacuum chambers required several new developments and led to a remarkable technology transfer.

## 6.4 The Radio-Frequency Accelerating System

Another example where an apparently well known technology had to be revolution-ized by new ideas or technologies is the high-frequency system for the acceleration and storage of the electrons and positrons. Two different innovations were developed for LEP 1 and LEP 2, respectively.

### 6.4.1 Copper Cavities

The accelerating cavities have two functions. First, they must accelerate electrons and positrons from the injection to the final energy; second, while circulating for many hours and producing collisions, the particles lose energy owing to synchrotron radiation and this has to be restored continuously. Because of the considerable radiation losses, high accelerating fields have to be produced in accelerating cavities to compensate for these losses. An accelerating cavity consists of several cells (Fig. 6.7) which are electromagnetic resonators in which electric fields oscillate in time (high radio frequencies). Their phase is controlled in such a way that the field points in the right direction to accelerate the particles at the moment when a particle bunch traverses a cell. The electric fields in neighbouring cells are in opposite directions and during the flight time of the particles from one cell to the next the field must change its direction in order for the particles to be accelerated in all cells. During each traversal of a cavity the particles get a push and since they are circulating almost with the velocity of light they traverse the cavities many times per second. In previous storage rings copper cavities fed by rf transmitters were used.

In the first phase of LEP copper cavities had to be used since no other technology was available at that time. Each cavity consisted of five cells with five gaps of accelerating fields (Fig. 6.7). To compensate for the radiation losses at an electron

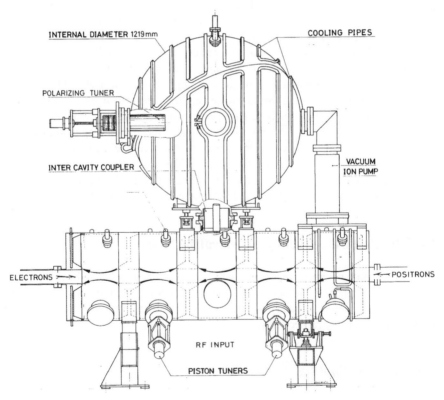

**Fig. 6.7** A copper accelerating cavity. *Lower part* with five accelerating cells, *upper part* storage cavity to reduce losses

**Fig. 6.8** Several copper units of the accelerating and storage cavity

energy of 50 GeV a total accelerating field of 400 MV along the ring circumference was necessary and in the first phase of LEP in total 128 such five-cell cavities were installed in the two straight sections 2 and 4 of the LEP ring. In Fig. 6.8 one such accelerating unit is shown. The fields in the cavities oscillated with a frequency of 352 MHz, which was chosen after long debates to minimize the operating cost. This relatively low frequency (usually 500 MHz was used in other colliders) was selected in view of the superconducting cavities to be installed for LEP 2 later. For this kind of cavity a lower frequency is more advantageous and it was planned to use the same power sources for both kinds of cavity.

To produce the required field strengths in copper cavities in the conventional way an enormous rf power would be needed. In contrast to normal accelerators, in which only a pulsed accelerating field for the duration of the acceleration is required (for linear accelerators of the order of microseconds, for synchrotrons about seconds or minutes), in a storage ring the compensation of the radiation losses must occur all the time while the particles are stored, usually 10–12 h. The fields in the cavities are produced by electric currents flowing in the walls of the cavities. Even for copper, which is a very good electric conductor, these currents produce considerable thermal losses and as a result an essential fraction of the rf power is not used to build up an accelerating field, but is lost. The electricity cost for the accelerating system turned out to be the dominant part of the operating cost of LEP even with its large radius chosen to minimize these expenses.

To make the operating cost acceptable an ingenious trick was used. Although the particles circulate all the time, they are not distributed uniformly around the circumference but are concentrated in a few bunches, normally four bunches of electrons and four bunches of positrons (circulating in opposite directions), each only a few centimetres long but separated by several kilometres. This implies that the accelerating field is needed only during the extremely short time during which the particles traverse an accelerating cavity, whereas for most of the remaining time the cavities are idle and use electricity for nothing. The laws of nature unfortunately dictate that a cavity with an accelerating gap has relatively large losses. However, spherical copper cavities can be constructed with relatively low heat losses in their walls. Advantage is taken of these facts in the following way. The cylindrical accelerating cavities are coupled with spherical low-loss cavities (see Fig. 6.7) in such a way that the electromagnetic fields reside in the accelerating cavities during the passage of particles but are transferred to the low-loss cavities in-between.[5] This scheme results in an overall saving of 40% of electricity, corresponding to several million Swiss francs per year. However, even with this scheme the total necessary rf power amounts to an impressive 16 MW.

To produce this large rf power at such high frequencies special emitters, called klystrons or magnetrons, are required. However, existing tubes were able to produce the extremely high power levels only in short pulses, typically for microseconds, whereas in LEP the cavities had to be fed continuously. What we wanted were klystrons with a nominal continuous rf power of 1 MW and an efficiency (output rf power/input ac power) of more than 60%. In total 16 klystrons would be installed initially each feeding four accelerating cavities. To allow access to the klystrons and their auxiliary equipment during operation they were installed in special tunnels running parallel to the main tunnel.

No such klystrons existed; however, the CERN expert thought that the development of devices with such parameters was within technical reach. Hence, we intended to stimulate industry to start such development, but we had to take into consideration a critical argument. Klystrons have an average lifetime of several thousand hours and must then be replaced. They are consumables. If we signed a development contract with only one firm, we would later depend exclusively on that firm. Because there would be no competition, it could ask any price for klystrons during later years of LEP operation. Hence, we decided to involve several firms in the klystron development, a technologically extremely demanding task. Only a few firms worldwide had the necessary know-how to be involved in such a venture. We signed development contracts with three firms, one in Japan and two in Europe. The Japanese firm declared after some time that it was not able to fulfil our expectations, whereas the European firms (Thomson in France and Phillips-Valvo in Germany) managed to meet our specifications after considerable efforts.

It should be mentioned that the copper accelerating system of LEP 1 with its final 128 cavities with a total length of close to 300 m, a continuous rf power of

---

[5] This corresponds to the well-known phenomenon of two coupled pendulums where the oscillation energy goes back and forth between the two pendulums.

16 MW and a total accelerating field of 400 MV represented one of the largest linear accelerators for electrons with a continuous beam, although it was 'concealed' in the LEP ring.

## 6.4.2 Superconducting Cavities

As explained in Chap. 2, the radiation losses increase very rapidly with the energy of the circulating particles and consequently a considerable amount of rf power has to be added to reach energies in the region of 100 GeV. To achieve this with copper structures would have been practically impossible. The losses in the walls of copper cavities are very large, with only about 10–20% of the rf energy being transmitted to the beam. The large heat losses require powerful cooling of the cavity walls and in practice accelerating fields are limited to about 1.5 MV/m. Hence, to reach energies in the region of 100 GeV with copper cavities was impractical because there would not have been sufficient space in the straight sections of the LEP ring, not to mention cost limitations. A new technology had to be employed, not yet available for the first phase of LEP, which, however, became sufficiently mature in time for the upgrading of LEP. For that purpose CERN had supported for several years the development of superconducting cavities.

The losses of electric currents can be reduced considerably in some metals by cooling them to temperatures below a 'critical' temperature, usually only a few degrees above absolute zero temperature. For a dc current the electric resistance can disappear completely (a phenomenon called 'superconductivity'), whereas for an oscillating current a residual resistance remains depending on the frequency, the temperature and the quality of the material. Nevertheless, for the frequencies used in accelerating cavities, large reductions of the losses, of the order of 100,000, can be obtained. Superconducting cavities also have the great advantage that much higher accelerating fields can be produced, thus reducing the required total length of the accelerating structure.

In Europe the development of superconducting cavities started when we built such cavities for the first time in the 1960s at the Research Centre of Karlsruhe (FZK). The oscillating currents do not penetrate deeply into the walls of a cavity and therefore it suffices to clad the cavity walls on the inside with a thin layer of the superconducting material. We started with copper cavities clad on the inside with a layer of lead, which is a metal that exhibits superconductivity at relatively high temperatures. Since lead is not very stable in open air and the cleanliness of the surfaces is quite important, we changed rather soon to cavities made of solid niobium, a metal which becomes superconducting at a critical temperatures of 9.20 K ($-263.8$°C). Such low temperatures can easily be produced by using liquid helium as a coolant, but of course the cavities have to be installed in cryostats. Niobium has become the favoured material since it has excellent properties for cavity manufacture. At first the cavities were manufactured from electronically welded niobium sheets, but later they were machined out of solid niobium and finally copper cavities with a niobium layer on their inside were used.

I invited Herbert Lengeler from CERN to come to Karlsruhe and lead the group developing the superconducting cavities, which he did very successfully for about 2 years. At a certain moment we also involved industry in the manufacturing of the niobium cavities. At Karlsruhe the development of superconducting cavities was continued by my colleague Anselm Citron. The problem in producing superconducting cavities lies in the requirement that not only the losses should be low, but in addition high accelerating fields must be possible without electric breakdown. This can be achieved only if the metal is very pure and the surface very smooth and clean. It took many years of great efforts to arrive at this goal. When Lengler returned to CERN a group for superconducting cavities was established there, directed later by Phillippe Bernard, and several other laboratories (e.g. DESY at Hamburg and Cornell University in the USA) became engaged as well in an active development programme.

In the 1970s the technology was still not sufficiently advanced that one could dare use it in a large project. Hence, the first practical application of superconducting cavities was not for the acceleration of particles but in a device called a 'particle separator'. Such separators are used to separate particles such as pions or kaons produced by protons hitting a target. For the first time a superconducting separator was built at CERN and was used in the 1970s at the Soviet Institute for High Energy Physics (IHEP) at Protvino. More than about 10 years was needed to advance the technology to such a state that its use for the storage of electrons in LEP could be considered.

At CERN the development of superconducting cavities for LEP started in 1979 and a number of prototype cavities had to be built before one could consider their production in larger quantities. To keep the losses small is more complicated in the case of high-frequency fields than in usual dc applications of superconductivity. At the frequency used in LEP, the penetration depth of the fields into the niobium is only about 100 atomic layers, i.e. less than 200 nm, and one is dealing essentially with surface effects, implying that the metal must be very clean and the surface smooth. By far the biggest challenge for superconducting cavities lies in the production of extremely clean cavity surfaces.

The first cavities were produced from high-purity niobium sheets developed in close cooperation with industry. Cavity half-cells were made by turning on a lathe and they were then joined by electron welding of extraordinarily high quality. Sophisticated diagnostic methods had to be developed for the testing and visualization of surface defects in these cavities. Encouraging results were obtained with massive niobium cavities. Losses were reduced by factors of about 100,000 and accelerating fields of 5–6 MV/m could be maintained, an impressive improvement compared with copper cavities with an accelerating field of 1.47 MV/m. One such unit is shown in Fig. 6.9. Nevertheless, it was decided to change to copper cavities clad on the inside with a layer of niobium which was deposited by sputtering or evaporation. A niobium thickness of 1–5 μm is sufficient because of the small penetration depths of the rf fields. Cavities of this type present several advantages. The heat conductivity of copper is about 10 times better than that of niobium, which ensures better thermal stabilization of any surface defects and thus higher accelerating fields

**Fig. 6.9** A superconducting cavity with four niobium accelerating cells

can be obtained. Another argument in favour of such cavities is the high cost of pure niobium, which represented about 25% of the total cost of cavities fabricated from massive niobium.

In practice, accelerating fields are limited by localized microscopic defects which lead to local heating and subsequent field breakdown (quench of superconductivity). Other defects act as emitters for electrons which are accelerated, hit the walls and cause additional losses. To prevent such defects, special installations had to be created for the chemical treatment, rinsing with ultrapure water and assembly under dust-free laminar airflow.

One issue which had to be considered at an early stage was the choice of the frequency of the accelerating fields. Careful optimization had to be carried out. With a decreasing rf frequency the rf losses decrease but, on the other hand, the dimensions of the cavities and hence their production cost increase. When a frequency of 352 MHz was chosen for the copper cavities, optimization of the superconducting cavities had already been taken into account. It was planned from the beginning that copper and superconducting cavities would be used in parallel and both would be fed by the same type of klystron.

Indeed a number of scenarios were presented at the time of the approval of LEP with various choices for the increase of the energy, first by adding superconducting cavities and finally replacing copper cavities by superconducting cavities to boost the energy to 100 GeV and beyond. For such an energy a total accelerating voltage of 4,000 MV would be necessary (10 times that for LEP 1 with 50 GeV) requiring an active length of the cavities of about 800 m, occupying completely two of the straight sections.

When the development work at CERN and partially carried out with industry had reached a certain state of success, the manufacture of the bulk of the

**RF voltage**

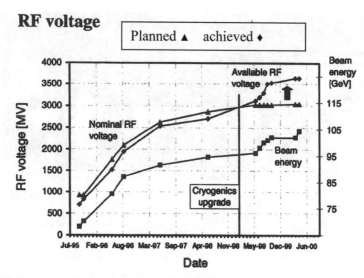

**Fig. 6.10** The progressive increase of the total accelerating voltage and the corresponding maximum energy increase. First period with copper cavities, second period (cryogenics upgrade) with superconducting cavities. The *bottom curve* shows the beam energy obtained (*right scale*)

superconducting cavities was contracted out to industry and over 4 years superconducting cavities were progressively installed in LEP (Fig. 6.10), boosting the beam energies first beyond what could be achieved with copper cavities and in a last dramatic run even to 104 GeV, the final step in terms of beam energy (see Chap. 14). On 26 February 1999 the last module containing four superconducting cavities was installed and in a little ceremony Enrico Chiaveri, Head of the Cavity Technology Group, congratulated his team on a difficult, but well-done job. In the final configuration LEP contained 288 superconducting cavities with an accelerating field of 7 MV/m, over 16% higher than the design figure.

The cooling of the rf cavities to liquid-helium temperatures required an extraordinary cooling plant, which was established under the leadership of Dietrich Güsewell. A particular effort was needed during the last stages of pushing the LEP energy to its highest values (see Chap. 14).

## 6.5 Other Components

### 6.5.1 Transport in Tunnel

The large size of LEP posed special problems for the transport in the tunnel. For the installation and operation of LEP it was crucial to transport both people and material in an efficient and fast way. Since only a few shafts were available for access, it usually happened that from the shaft to the working place several kilometres had to

**Fig. 6.11** Monorail train for the transport of bending dipole magnets

**Fig. 6.12** Monorail train for workers

be covered, implying that a non-negligible fraction of the working time would have been used for walking! Of course, when workers arrived at their work location the necessary material for installation or maintenance had to be in place.

After the available space in the tunnel, the frequency of transport and security issues had been studied, a monorail train suspended from the ceiling was chosen. Different convoys of trains could be assembled. Each consisted of two cabins for the driver, one at each end, one or two motor elements each capable of pulling 7.5 tons; a special vehicle for braking and finally different wagons depending on the special need. For example, there were wagons each accommodating eight people (Fig. 6.11) and special wagons for the transport of the dipole magnets (Fig. 6.12). Convoys could be up to 52 m long and move with a speed of 100-200 m/min depending on the load. Of course, a strict time schedule for the different trains had to be established and this was controlled by computers. This transport system turned out to be very successful and allowed 60,000 tons of material to be put in place within only 18 months.

## 6.5.2 Control System

A facility such as LEP is much too complicated to be controlled 'by hand'. It has to give to the operators what has been described as an 'extended arm and eye'. It must provide the ability to set and switch thousands of pieces of equipment distributed over many kilometres and observe the results as if they were in the same room. Controlling individual pieces separately would be impossible and hence the control system must be turned into a computer control system which provides the ability to set up the whole accelerator for different operating requirements by using a computer model to determine the optimum settings for the individual components. Of course, the system must also record any malfunctions and warn the operator of any event that might require his intervention.

To save money it was decided to make the LEP control system compatible with the old Super Proton Synchrotron (SPS) control system designed in the early 1970s and to operate LEP from the enlarged SPS control room on the Prévessin site of CERN. Since microprocessors had become available, a fully distributed, multicomputer system could be developed consisting of 160 computers and 2,000 microprocessors in 24 underground areas and 18 surface buildings. Such a system needs a secure and rapid means of transmitting signals and for that purpose a collaborative programme with IBM was started. IBM was developing the Token Ring local area network and before it was released to the public it was tested out at LEP. Here is another example of technology transfer. Industry likes to use CERN as a test bed for new developments since CERN as a user is very intelligent and can help to find errors. However, the IBM cables were not designed for the large distances at LEP and the time-division multiplex system usually used for telephone trunk lines was adopted for LEP purposes by installing two coaxial cables around the LEP tunnel. These cables were also used for other purposes (e.g. channels for experimental teams) since the cost of installing just one cable around the ring was about CHF 150,000. Optical fibres were utilized whenever possible; however, they are quite sensitive to radiation damage and hence could not be used in the tunnel.

The CERN director-general very rarely intervenes in decisions concerning special technical issues. When I did it exceptionally, most of the time I thought it was reasonable, but once I definitely made a mistake. The computers distributed around the ring consisted of crates containing a number of microprocessor modules. Originally the development was supposed to be done at CERN, but in line with our general management policy I insisted that work which could be done by industry should be done there. Hence, at my request a contract was given to a competent French firm. To our surprise the firm cancelled the contract after some time because it had hoped to get a similar contract for military purposes, but this did not happen. This created a serious problem since these modules were critical for the installation of LEP. A crash programme had to be started at CERN and thanks to the great effort of Pier Giorgio Innocenti and his collaborators a delay in the commissioning of LEP could be avoided.

### 6.5.3 Conventional Equipment and Safety

Apart from those components of LEP which required innovative solutions several systems had to be provided for which conventional components available 'off the shelf' could be used. However, because of the size of LEP their employment required great efforts and a large fraction of the total budget of LEP went into these parts of the project. Here they are mentioned only briefly.

A large number of machine components required precisely stabilized dc power with power levels ranging from 7 MW down to fractions of a kilowatt. For the magnet system alone 757 power converters were needed and 580 were needed for the vacuum ion pumps.

Powerful cooling and air-conditioning systems had to be provided. The heat generated by the components in the tunnel was removed by a hydraulic system of chilled water transporting the heat to cooling towers at the surface. This was done by a complicated distribution system involving many hundreds of pumps controlled by computers. An installation for air treatment and distribution supplied and extracted air to ensure a sufficient change rate in the different underground areas both for the normal operating conditions as well as during emergencies. The 134 air-treatment and 86 air/smoke extraction units were managed automatically by 136 microprocessors.

In full operation LEP needed electrical power of 75 MW when running at 51 GeV and 160 MW when running at 100 GeV. An extensive supply and distribution system had to be created with transformers, cabling and controls. In total more than 50,000 cables had to be installed with a total length of 4,700 km.

Radiation safety and access of personnel to the tunnel were an important part of the whole project. Monitors installed in areas accessible during operation were the backbone of personnel radiation protection. Radiation levels were relatively high in the curved sections of the ring and immeasurably low at the ground surface. Of course, the radiation around CERN was continuously monitored and controlled by the Swiss and French authorities (see Sect. 5.2). During the whole existence of LEP no incident was recorded. Access to the tunnel was by 20 doors which were electrically locked during operation. When LEP was switched off temporarily, access was permitted by magnetically coded access cards. The name of the person entering was recorded by the operator, who released one of the interlock keys from a key-bank. This key had to be carried all the time by the person who was in the tunnel. Only when all the keys had been put back into the bank could LEP be started again. This rigorous procedure was unavoidable since it took quite a long time to search the long tunnel for anybody still in it.

## 6.6 Injection System

When one wants to put a load into space, one cannot do it in a single step, but several rockets have to be fired one after the other. In a similar way, one cannot accelerate particles in one go to very high energies; one has to do it in stages. The difference is that with rockets one starts with the big one, followed by smaller ones, whereas particles are accelerated first in a small device and are then transferred to bigger ones. Hence, electrons and positrons had to pass through a whole chain of devices before they were injected into the LEP main ring.

As mentioned in Chap. 3, the use of the existing accelerators as injectors for LEP reduced the cost of the project. However, it was not only a financial issue. The whole injection chain is a very complicated system and its reliable operation benefited from the experience and competence which had been accumulated at CERN over several decades.

Before particles are accelerated, they must be produced. This is not a problem for electrons. Like in old TV sets, they 'evaporate' from a hot wire. A much more difficult issue concerns the production of a sufficient number of positrons. Positrons are antimatter particles and do not normally exist in nature. According to the fundamental laws of physics they can only be produced in pairs of electrons and positrons if sufficient energy is made available. This can be achieved most easily by energetic electrons hitting a heavy-metal target, where they are decelerated and produce bremsstrahlung (hard photons with much more energy than those in a usual X-ray device are needed), which then in the same target produces electron–positron pairs. It is easy to separate the positrons from the electrons using magnetic fields. However, the number of positrons produced in this way is rather small and they have to be accumulated before they can be accelerated further. For that purpose a special little accumulator ring, Electron–Positron Accumulator (EPA), had to be built.

The total preinjector assembly works in the following way (see Fig. 6.13). An intense beam of electrons is accelerated in a linear accelerator to 200 MeV and hits a

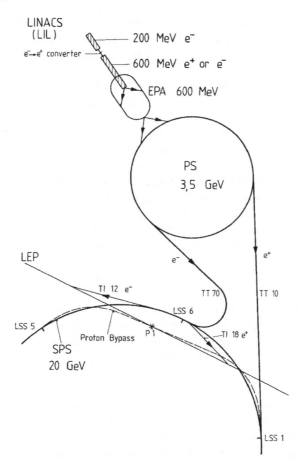

**Fig. 6.13** Injector system of LEP using the existing accelerators at CERN, the Proton Synchrotron (*PS*) and the Super Proton Synchrotron (*SPS*). The LEP Injector Linac (*LIL*) and the Electron–Positron Accumulator (*EPA*) had to be built. On the scale of this figure the LEP ring appears as a straight line

tantalum target for the production of positrons. Both electrons and positrons are then accelerated in a second linear accelerator, LEP Injector Linac (LIL), to 600 MeV. The performance the two linear accelerators had to meet was not a trivial issue and a considerable development effort had to be undertaken. Since the resources at CERN were limited, a contract was signed with LAL, the French Accelerator Laboratory at Orsay near Paris. Over 2 years a concentrated and successful study and technical work were carried out there, building and testing various prototypes.

After electrons and positrons have been accelerated in the linear accelerators they are transferred to the electron accumulator ring EPA. To accumulate there a sufficient number of positrons in eight equidistant packets ('bunches') about 11 s is required, whereas for the more abundant electrons only 1 s is needed. To achieve a fast switching from electron to positron operation, the directions of the fields in the magnets are kept constant, which implies that electrons and positrons have to be injected in different directions to make them circulate in the opposite sense. Normally two cycles of positron accumulation in EPA are followed immediately by two electron cycles.

As a next step the particles are transferred to the old proton accelerator Proton Synchrotron (PS). The circumference of EPA was chosen to be 125.665 m, which is exactly one fifth of the PS circumference and provides an optimum cog-wheeling between the two machines and allows the transfer of the eight EPA bunches into eight bunches during one revolution in the PS.

To facilitate the transfer of the particles from the preinjector system to the PS, the two linear accelerators and EPA had to be located as close to the PS as possible. Unfortunately the only space available was an intersection of two of the main roads of CERN and a car park (see Sect. 4.8). In order not to close the roads completely the overall length of the preinjector chain was minimized by pushing LIL partly into EPA (Fig. 6.13), injecting the particles into the accumulation ring from the inside. To save space the klystrons which fed the linear accelerators (the 'linacs') were mounted on a floor on top of the linear accelerators themselves.

The particles with an energy of 600 MeV and properly bunched were taken from EPA to the PS and after having been accelerated to 3.5 GeV they were transferred to the SPS. There electrons and positrons were each accumulated in four bunches and taken to 20 GeV before being injected into the LEP main ring.

The PS and the SPS were constructed for acceleration of protons to much higher energies (27 and 400 GeV, respectively), yet because of the synchrotron radiation emitted by the electrons increasing very rapidly with energy, the maximum energies for electrons had to be limited to much lower levels. With this restriction it turned out that surprisingly modest modifications had to be made to the PS and the SPS. The main additions were rf cavities for the acceleration of electrons or positrons in both machines. Of course, it is easy to say that only a few additions were necessary. Actually an enormous amount of work had to be done to cope with all the challenges regarding beam behaviour, ejection and injection. Here the competence of the CERN staff and their long experience were essential for the final success.

Because of the relatively small changes, the original modes of operation of the PS could continue and it became the most versatile accelerator in the world. It could accelerate protons, antiprotons, ions, electrons and positrons in interleaved cycles, each lasting a few seconds. Thus, experiments at LEP could be served at the same time as other experiments at the other facilities.

## 6.7  The Final Steps

### 6.7.1  Installation and Dismantling

To install 60,000 tons of equipment in the tunnel with extremely limited access through a few narrow shafts was both a technical and an organizational challenge. To recuperate some of the time lost owing to the delays in excavating the tunnel, the installation team embarked on a crash programme in September 1987, after the

**Fig. 6.14** The first magnet was installed in the LEP tunnel by Prime Minister Jacques Chirac of France and President Pierre Aubert of Switzerland on 4 June 1987. From the *left*: ?, Günther Plass, Henry Laporte, Gerard Bachy, Jacques Chirac, Pierre Aubert, Herwig Schopper, French Minister of Research Jacques Valade (partly covered)

first magnet had been ceremonially installed in the tunnel in the presence of Prime Minister Jacques Chirac of France and President Pierre Aubert of Switzerland on 4 June 1987 (Fig. 6.14). As soon as part of the tunnel had been vacated by the civil engineering crews, the installation of components started. The problem of transporting the elements to their final locations and of getting the workforce to the proper places in time has been mentioned already.

The installation of some of the components was a challenge in itself. To place the heavy magnets with a precision of a fraction of a millimetre in the tight confines of the tunnel required the development of special manipulators, which were affectionately named 'lobster' and 'crayfish' because of their arms resembling somewhat those of the corresponding animals. In only 2 years the enormous job of installing the components of the main ring, connecting the different parts and putting in place all its infrastructure, such as cooling and ventilation, electrical power and control systems, was achieved. Figure 6.15 shows a view of the LEP tunnel after the installations had been completed.

## 6.7.2 Dismantling

Even after LEP was shut down in 2000 its dismantling presented a technical challenge. It had to be done safely and quickly since the LHC components were already

**Fig. 6.15** View of the tunnel with LEP fully installed

waiting to be installed in the LEP tunnel. Within 14 months 30,000 tons of material (4 times the weight of the Eiffel Tower) was removed from the tunnel through a few narrow shafts. Many items of equipment still in working order had been given to scientific institutes all over the world. About half of the total material was recycled. For example, some of the yokes of the dipole magnets consisting mainly of concrete were cut into halves and used to strengthen building materials and for roadwork. Another 10,000 tons of material from the experiments had to be removed. Only the huge magnet of L3 remained in place since it was to be used for the ALICE detector at the LHC.

However, even the smallest box of nuts and bolts had to be checked for radioactivity before it left the tunnel. Fortunately even after 11 years of operation only 2% of the material had to be classified as 'very slightly radioactive' from having been in contact with the beams. This is an advantage of an accelerator for electrons that extremely low levels of radioactivity are produced.

## 6.8 The First Collisions

The commissioning of a large facility is always an exiting experience since it has to be proven that no major errors were made in its design and that all components work as expected. In addition, for a machine such as LEP, an adventure into unknown territory, unforeseen effects might appear which were previously unknown and which might impair the proper functioning of the facility.

To speed up the commissioning of LEP we did not wait until the installation of the whole main ring had finished. As soon as the first octant was ready, it was tested. On 12 July 1988 at 23:43 LEP lost its virginity when a beam of four positron bunches with an energy of 18 GeV was successfully transported along the complicated transfer line connecting the SPS to LEP (winding in three dimensions since the vertical levels of the SPS and LEP differed significantly), injected into LEP and guided 3 km to the end of the first completed octant. Usually at such first tests trivial problems occur, such as a magnet having been connected with the wrong polarity or it even happened that a beer bottle had been left in the path of the particles. No such calamity was observed and to everybody's satisfaction the beam glided through the octant without even the necessity of using correction elements available to adjust the beam orbit. This perfect start-up was permitted by the faultless turn-on of the whole injection chain (LIL, EPA, PS and SPS and the necessary transfer lines), a great triumph of engineering. This first test could even be performed in 'interleaved mode' when the PS and the SPS were continuing their proton programme without interruption.

Another milestone followed 1 year later when the installation of the whole ring had been achieved. On 14 July 1989 at around 15:00 the control room was packed to the doors when the accelerator physicists, the engineers and the technicians anxiously watched the first attempt to coax a beam around the 27 km of LEP. Within 1 h of tuning the machine, a beam of positrons was steered around the ring to complete the 'first turn'. This moment is always crucial in the commissioning of any accelerator since it shows that there was no major fault in the design of the facility and that the thousands of components were doing their job . The control room became a scene of celebration and several bottles of champagne were opened (Fig. 6.16). Particularly our French colleagues enjoyed this happy moment since it happened on the day marking the bicentenary of the French Revolution. Many scientists and engineers were involved in getting the machine started, but only two can be mentioned here, Steve Meyers and Albert Hofman.

The celebrations of the first turn had to be followed by hard work bringing progressively into action all the systems needed for acceleration and collisions of the beams. The rf copper cavities for the acceleration were activated and the timing of their fields was adjusted so that incoming bunches of particles would arrive at the right time to be 'captured' and accelerated. This soon resulted in several hundred turns of the beam and a beam lifetime of 25 min. Measuring the revolution frequency and knowing the speed of light allowed the circumference of the beam orbit to be calculated precisely and it showed that the 27 km agreed with the design value to an accuracy of better than 1 cm, another triumph of precision engineering!

In the course of these preparations a mysterious effect appeared which could not be explained for some time. It turned out that the oscillations of the particles in the vertical and horizontal directions were coupled in a strange way and this would have limited the performance of the machine. It took quite a while to find out that the major source of this mysterious behaviour was the thin nickel layer which was used to improve the bonding between the aluminium and the lead of the vacuum chamber (see Sect. 6.3). Nickel is magnetic and any magnetic material can distort the very

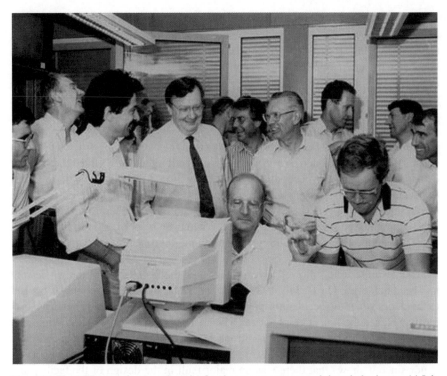

**Fig. 6.16** The LEP control room when the first beam was sent around the whole ring on 14 July 1989. From the *left*: ?, Diether Blechschmidt, ?, Carlo Rubbia (then director-general), Manfred Buehler-Broglin, Donatus Degele (sitting), Herwig Schopper, ?, Steve Myers

precise magnetic fields which guide the particles. The focussing quadrupoles and so-called skew quadrupoles had to be adjusted to counter this problem. Eventually the nickel layer was demagnetized.

On 25 July 1989 electrons were successfully injected for the first time and a beam tuning similar to that for the positrons was carried out with the aim of storing as many particles as possible. By early August the beam currents had reached an intensity of about $500\,\mu A$. Finally, on 12 August electrons and positrons were circulating in opposite directions at the injection energy of 20 GeV. The rf accelerating system was turned on to compensate for the synchrotron radiation losses so that the beams continued to circulate at this energy and the intensity was built up by injecting more pulses. Finally, both beams were taken to an energy of about 46 GeV and all was ready for the first collisions. Beam collisions have to be avoided during the acceleration and to achieve this the two beams are kept apart by special electrostatic separator plates. When the electric fields in these separators are switched off, the beams are expected to collide in the centres of the four large detectors. However, it turned out that the beam lifetimes were poor and the machine performance was not good enough to cause the beams to collide under the prevailing conditions.

More tuning had to be done and finally on 13 August 1989 everything was working properly and late in the evening the separator plates were switched off. This moment had been eagerly awaited by the hundreds of physicists who had prepared the LEP experiments. In the control rooms for the experiments many eyes were fixed on the computer screens showing the registered events. Ten anxious minutes later, at 23:13 the display lit up in the OPAL control room and the first event bearing the marks of a Z particle was visible. By the time a printed copy of the event had been brought from OPAL to the LEP control room, the telephone had been ringing again and a further five events from the other detectors had been reported. LEP physics was under way and a new era in experimental particle physics had started!

The first real physics run started on 20 September, with the machine luminosity rising to about one third of its design value in the 3 months until the end 1989. This fantastic performance allowed the four detectors to register some 17 million Z particles and some 40,000 W particles during the first 5 years of operation. This opened up a new domain of precision experiments. When Carlo Rubbia went to Stockholm to receive the Nobel Prize for the discovery of the W and Z particles his experiment had observed half a dozen Z particle events.

Thus, from the point of view of technology, LEP can be considered as one of the most successful large projects, a success which could be achieved thanks to the competence and devotion of many staff members at all levels. At the end of the project the project leader, Emilio Picasso, gave me a transparency which read:

*Six stages of a project:*

- *Wild enthusiasm*
- *Total confusion*
- *Complete disillusion*
- *Search for the guilty*
- *Punishment of the innocent*
- *Promotion of the non-participants.*

We made every effort to realize LEP in a fair manner; however, human shortcomings cannot be completely avoided in such a large venture. Those who feel that they were treated in an unjust way might get some consolation from the above list, which seems to hold true for all kinds of projects.

# References

1. CERN (1983) LEP design report, vol I. The LEP injector chain. CERN-LEP/TH/83-29, CERN (1983) The LEP injector chain CERN/PS/DL/83-31, CERN (1983) The LEP injector chain CERN/SPS/83-26, LAL/RT/83-09, June 1983. CERN, Geneva
2. CERN (1984) LEP design report, vol II. The LEP main ring. CERN/LEP/84-01, June 1984. CERN, Geneva
3. Wyss C (ed) (1996) LEP design report, vol III. LEP2 CERN-AC/96-01 (LEP2). CERN, Geneva

# Chapter 7
# The LEP Experiments –
# Institutions in Themselves

A remark concerning the nomenclature may seem appropriate. Very often one speaks of 'experiments' at the large accelerators. However, the detectors to observe the collision events are extremely complicated facilities which have lifetimes of more than 10 years, cost on the order of EUR 100 million and are constructed and exploited by international collaborations which change with time. They have an internal organization with a certain hierarchy, committees and their own financial resources. The cooperations to design, build and use these detectors can be considered as a realization of 'distributed' international institutes. In other fields the creation of such distributed functional entities has and is being discussed. In elementary particle physics they were so badly needed that all organizational or social problems had to be surmounted. The study of the history of these large collaborations could be useful for other fields. For LEP the situation was particularly vital since the participating scientists were spread all over the world and for the functioning of such 'institutes' new ways had to be explored. Communication was an essential item and the World Wide Web was invented and developed originally for this purpose.

## 7.1 The Approval of the LEP Detectors

A facility such as LEP is realized with concrete expectations hoping that the mysteries of the microcosm can be unveiled somewhat more, with guidance coming from theory. The art of designing a detector has to cope with two aspects. On one hand, one wants to clarify specific questions; on the other hand, the detector has to be sufficiently versatile not to miss any unforeseen phenomena. To achieve these goals for a unique machine such as LEP the obligation would be to involve the most brilliant minds in the world and to benefit from their imagination and experience. This requires a period of consultations, discussions and deliberations and meetings of future potential users are an important element during this procedure. Finally the advice of consultative committees is required before decisions can be taken.

H. Schopper, *LEP – The Lord of the Collider Rings at CERN 1980–2000*,
DOI 10.1007/978-3-540-89301-1_7, © Springer-Verlag Berlin Heidelberg 2009

## 7.1.1 A Meeting in the Swiss Alps and Letters of Intent

In June 1980 a first meeting of prospective users took place in Uppsala, Sweden, followed by other workshops organized by the European Committee for Future Accelerators (ECFA). The most important one was a meeting at Villars-sur-Ollon [1] in the Swiss Alps in June 1981 chaired by John Mulvey, Chairman of ECFA. This meeting took place well before the approval of LEP. Reviews of possible physics investigations at LEP, of the status of the design of the project and of the various experimental techniques were given. The main motivation for holding this meeting was, however, to bring scientists together, not only from Europe, but also from the whole world, to establish personal relations which would eventually lead to cooperations. Such a 'marriage market', as I called it, was essential in view of the uniqueness of LEP and of the limited number of detectors. Indeed many physicists came and benefited from this opportunity. The meeting took place in a hotel that was managed by Club Méditerrané and the *gentil animateurs* were seriously disappointed when the physicists spent their time sitting at the bar and discussing physics instead of joining them for various types of entertainment. Hans Bøggild from Denmark composed the following poem during the meeting:

> There once was a place called Villars,
> A palace with more than one star,
> They talked about LEP,
> The future of HEP[1],
> But decisions were made at the bar.

The question of how to organize big collaborations and their sociological aspects were main issues to be considered. The large UA1 experiment at the CERN Super Proton Synchrotron (SPS) collider was in some ways a precursor of the future LEP collaborations, but a new level of sophistication had to be achieved. The LEP detectors, each with some hundreds of scientists, were a step to the even larger collaborations at the LHC with more than 1,000 participants.

My aspiration was to enable as many scientists as possible to do research at LEP. For that purpose I pointed out that detectors should be considered as facilities and different experiments using the same facility might be distinguished, e.g. by using particular triggers (see Sect. 8.2.3), by adding special additional equipment or by a new idea for the analysis of the data. I expressed my opinion that

> A large collaboration should not be a unique block, like a single big ship with one captain, one big crew and one unique common course which all have to follow blindly. I would rather consider it as a convoy of smaller ships which have come together for a certain task (building the facility) but which has a large flexibility and well-determined subunits. There can be small and big ships in the same convoy with specialized techniques (steam or diesel engines or even sailing boats). Sometimes a small but powerful tug can drag along a big immobile ship or, inversely, the whole convoy can help a small sailing ship during a calm to

---

[1] HEP stands for high-energy physics.

get along. Of course, there are several captains and, as time goes by, some boats may leave this particular convoy and join another.[2]

At this meeting a rather broad consensus was achieved on the further procedure for selecting and approving LEP detectors:

- Only four interaction regions should be used for detectors, not only because of financial restrictions; two others could be opened later if there were new ideas. Indeed they were only used for the LHC experiments ATLAS and CMS.
- Letters of intent were to be submitted and they would be evaluated by the LEP Experiments Committee to be set up in 1982.
- For proposals to be approved it would be necessary for them to be scientifically valuable, technically feasible and financially possible, and the collaboration should be competent and sufficiently strong.

The first approvals were expected for the middle of 1983 after the general scenario for the first generation of LEP experiments had been determined, leaving 4 years for the construction – a rather short period. Various scenarios were considered, such as three standard detectors with one small specialized detector, only one or two all-purpose detectors and the use of an existing detector.

Some concern was expressed in view of the participation of scientists from smaller member states. Because of the uniqueness of LEP it was also to be expected that groups from non-member States would want to use LEP. I proposed, supported by ECFA, that participation of groups from non-member states should be judged on the basis of merit and feasibility. The sharing of cost and efforts should be settled on a case-by-case basis, assuming that the participation of member state groups in any detector would be natural and hence automatic.

I also raised a problem of some sociological importance. It has become practice that the publications of large collaborations carry the names of all participants. Many dozens or even hundreds of authors' names make it very difficult to recognize and appreciate an individual contribution, which has negative effects for career advancements or personal ambitions and pride. I suggested that one might consider the option that a large collaboration publishes several common papers on the design and construction of the detector ('facility papers'), whereas the results of particular 'runs' (taking data under particular conditions, e.g. triggers) or specific data analysis ('experiments') could be assigned only to those who had contributed to this specific activity. An alternative way to express the different contributions could be to order the authors' names according to their involvement instead of in alphabetical order. I completely failed to gain acceptance of these proposals – the problem is still with us today!

In parallel to the ongoing approval procedure for the LEP machine, the evaluation of the proposals for detectors continued. In total six letters of intent for detectors

---

[2] The LHC collaborations have become so large that only the 'convoy model' seems to work.

were submitted and they became known under the acronyms ALEPH, DELPHI, L3, OPAL, ELECTRA and LOGIC. The approved detectors will be described in more detail in Sect. 8.2.2.

## 7.1.2 The LEP Experiments Committee

The LEP Experiments Committee chaired by Günter Wolf from DESY with 14 members was established and began its work in March 1982. Its task was to evaluate the proposals and provide advice for a final decision. In monthly meetings it evaluated very carefully the six proposals, taking into account the physics relevance, the technical feasibility, the competence and strength of the groups and it also tried to verify the cost estimates. The committee did an excellent job and in a closed meeting on 13–14 July 1982 it formulated its final conclusions, stating that all proposals were physically well founded and technical feasible; hence, there was no reason to refuse of any of the proposals. However, a very slight preference was given to OPAL relative to ELECTRA, which followed the concept of the TASSO detector at PETRA. DELPHI was given some priority over LOGIC since it was characterized by a more open geometry offering some advantages for measuring particle momenta. LOGIC had been predominantly proposed by American groups, but this was not the reason for giving it less priority. The committee suggested some changes to the proposals to make the detectors more complementary.

In an earlier public meeting of the committee I had warned that the final judgment would not be based on a strictly individual evaluation of the proposals, but that an overall balance had to be found with respect to technical diversity and risks, distribution among member states and non-member states and last but not least the financial situation. Some observers from the media wondered how CERN would react to the political pressures to give preference to scientists from the member states against scientific objectivity, when not all proposals could be accepted and a "lot of careers are at stake" [2]. One colleague remarked that "the committee looks at the science and the Director-General looks at politics." Of course, the final decision had to take into account all aspects – scientific, financial and political ones. In particular, I wanted to avoid a procedure which in the USA is called a 'shoot-out', namely simply rejecting some of the proposals. Therefore, when the LEP Experiments Committee informed me that it had given preference to four of the proposals I insisted that negotiations should start between the four favoured collaborations and the other two with the aim that all scientists who wished to join the accepted proposals could do so by a kind of marriage between the groups. Indeed this was achieved to a great extent and was facilitated by the fact that the approved detectors needed additional manpower and financial resources. However, a colleague who was bitter that his proposal had not been accepted told me that "the difference between marriage and rape is sometimes only a technical one."

## 7.1.3 The Conditional Approval

Starting on the morning of 15 July 1982 I invited on each consecutive day the representatives of one group and informed them that a conditional approval could be given to their proposal, subject to certain changes they would have to implement. ALEPH was led by Jack Steinberger, a prominent physicist and Nobel Prize laureate. It consisted of 24 European goups, one American group and one Chinese group. OPAL, with Aldo Michelini as spokesperson (later followed by Rolf-Dieter Heuer), was supported by 20 institutions, including one from Canada, one from the USA, one from Israel and one from Japan. DELPHI, with Ugo Amaldi as spokesman, was supported by 37 European institutions. The L3 proposal directed by Samuel Ting, an outstanding scientist and Nobel Prize laureate, was supported by 35 institutions, including nine American institutions, two Chinese institutions and a large Russian group.

The LEP Experiments Committee had asked some questions and required changes. For ALEPH it was not clear whether sufficient space would be available in the underground caverns. For DELPHI, which had originally been proposed as a universal detector, more emphasis was required to be put on new Cerenkov counter techniques for particle identification. L3 was approved under the condition that the detector size be reduced by 10% so that the huge magnet would fit in an underground experimental hall. All collaborations were asked to clarify questions concerning the available personnel and financial resources. Within a few months the modified proposals were submitted and reevaluated by the committee, and the conditional approval was given on 18 November 1982.

After the negotiations with the groups whose proposals had been rejected and after the final design of the detectors had been achieved, additional institutions joined the collaborations. At the time of the approval we estimated that close to 1,000 scientists would be involved in the four collaborations. Close to 700 physicists came from member states, about 80 from CERN and nearly 200 from non-member states. During LEP's operation it attracted more scientists, and when it was closed down in 2000 more than 2,000 physicists were involved in LEP experiments.

With two general-purpose detectors (ALEPH and OPAL), one employing more advanced but riskier technology and the other based on safer technology, and two more specialized instruments (L3 specialized for photon calorimetry and DELPHI for hadron identification), the programme seemed to be well balanced.

At the beginning I watched the sociological behaviour of the various groups with some concern. Two collaborations (ALEPH and L3) were guided by very strong personalities (indeed Nobel Prize laureates), whereas the two others (DELPHI and OPAL) were organized rather democratically. To my surprise both modes of cooperation worked quite successfully.

When I was at DESY I liked the tradition of using nice acronyms for machines and detectors which are easy to remember, such as DORIS, PETRA and PIA. At CERN, in contrast, names of experiments were represented by simple logos, such as NA31 or UA1, indicating the north area or the underground area and the number of the experiment. Even with the power of director-general I did not succeed in changing this tradition for most of the experiments at CERN. However, for the LEP experiments I insisted on nice acronyms which could be easily remembered and which may evoke even pleasant emotions. Acronyms are usually abbreviations for long names which, of course, are immediately forgotten. Sometimes people wondered why L3 did not follow this request. When I asked Samuel Ting to propose an attractive name he suggested SAM. When I replied that it was not very elegant to use one's first name for the detector he replied that SAM had nothing to do with his first name but stood for 'Schopper approves me'. I refused to accept tis acronym and asked him to come back with another proposal. A few weeks later he proposed Magellan, the name of the fifteenth century Portuguese explorer. I found this to be excellent since it symbolized the worldwide participation of the L3 collaboration. However, a short time later Ting came back and informed me that the collaboration was against that name. Magellan had been shipwrecked after discovering the Philippines and had been slain by the natives – a bad omen for an international collaboration! At that moment I gave up and the collaboration was named L3 since the proposal had been submitted as the third letter of intent.

Some colleagues argued that the four approved detectors were too similar in some aspects and not all of them should have been approved. However, I thought that I had to resist such criticism for various reasons. As mentioned above, the two general-purpose detectors used different technologies with different risks. That both would be ready at the turn-on of LEP was hard to foresee, but in the end this was achieved thanks to the competence and enthusiasm of the scientists and technicians involved. This was also true for the other two detectors. Nevertheless, to have four detectors seemed justified also by the fact that no other facility like LEP existed in the world and hence a cross-check of the results obtained was only possible by different experiments at LEP. In addition, the total number of observed events would become larger by combining the results of several detectors and would thus improve the statistical significance in the case of rare events. Indeed these arguments turned out to be particularly relevant towards the end of the operation of LEP when the search for the Higgs particle took place. Last but not least, because of the great enthusiasm and interest among physicists I thought that this unique facility should be open to as many scientists as possible.

The next problem was in which underground caverns the four detectors should be located (see Fig. 7.1). Technical arguments helped to make a decision in favour of L3 and OPAL. L3 was the largest detector and needed an especially high hall, which for geological reasons could be realized best in area 2. OPAL, with a conventional magnetic coil, needed a lot of power and this could be provided in area 6 by the electricity network. However, we could not find any criteria for making a decision concerning ALEPH and DELPHI. Both of the relevant collaborations preferred underground area 8, which is much closer to the main laboratory than area 4 at the foot of the Jura Mountains. To solve the problem I invited the two spokesmen, Jack Steinberger and Ugo Amaldi, to a Directorate meeting and since after a long discussion we could not come to a conclusion, I suggested flipping a coin. This was accepted, a two Swiss franc coin was flipped and ALEPH was 'condemned' to go to

**Fig. 7.1** Aerial wiew of LEP showing the locations of the four LEP experiments (the present LHC experiments are indicated in *italics*)

area 4, whereas the experimental hall for DELPHI could be built in area 8. It has a large window looking like a Greek $\Omega$ which can be seen when landing or taking of from Geneva airport. Steinberger, being a very careful experimenter, asked for the coin and checked whether it really had two different sides! In the end it turned out that both collaborations were quite happy with this decision.

Although a large part of the four detectors was built outside CERN, the responsibility for coordinating the construction of the detectors (involving in the end about 2,000 physicists and engineers from 12 member states and 22 non-member states) and their installations in the underground caverns remained with the CERN management and presented a major challenge. It always seemed to me like a little miracle that in the end all the parts of this very sophisticated equipment arriving from all corners of the world fitted together and worked. Building these huge detectors within an extremely tight time schedule was a great challenge. The success was due to the competence and dedication of many people, above all to the directors for the LEP experiments, Erwin Gabathuler followed by Ian Butterworth, and to Horst Wenninger, who had been appointed Technical Coordinator for the LEP Detectors, and Franco Bonaudi, who was responsible for the infrastructure of the experiments.

## 7.1.4  A Typical Detector and Detection Methods

The challenges presented by the design, construction and use of the LEP detectors were considerable but different from those presented by the construction of LEP itself. Compared with those in previous facilities, the LEP detectors had to be able to cope with the higher energies at LEP, but above all a much higher precision was aimed at than ever before. It is not the purpose of this book to give an extended survey of the detector development and I must limit myself to a short review.[3]

The LEP detectors are assemblies of various components of an impressive size with a total weight of thousands of tons and are many metres high and long. Why is it necessary to build such huge detectors? Is it a kind of megalomania of physicists? Before this question can be answered, the various methods of detecting particles and determining their properties, such as energy, direction of emission and identity, have to be considered.

The designs of most of the detectors for colliders have some common features. Since the particles produced in the collisions are emitted in all directions, the detectors have to try to capture them in as large an angular region as possible.[4] This property is called 'hermiticity' and its importance had been recognized by the UA1 detector which discovered the W and Z particles in proton–antiproton collisions at CERN in 1983. To reconstruct an event completely, good hermiticity is important for the following reason. Some particles might leave the detector without any trace. This is the case for neutrinos, which are produced quite often in processes involving the weak interaction. Even more exiting is the search for some hypothetical particles such as the supersymmetric particles predicted by some theories. Although such 'traceless' particles cannot be detected directly, their existence and properties can be inferred indirectly from missing quantities, e.g. missing energy, momentum and other quantities.[5] To be able to exploit these conservation laws all detectable particles should be measured and for that purpose the interaction region should be completely enclosed and all 'cracks' between detector elements through which emerging particles might escape should be avoided.[6]

---

[3] For a complete review of modern particle detectors, see [3].

[4] The centre of gravity of the colliding particles is at rest in the laboratory and therefore the collisions are symmetric in the laboratory.

[5] According to the conservation laws of energy and momentum, for all the particles produced in the collision the total energy after the collision must be equal to twice the beam energy and the momenta must add up to zero.

[6] In some experiments looking for specific phenomena the detector observes particles emitted only into a restricted angular range. This is the case for the LHCb experiment.

In principle the detectors are structured like onions surrounding the interaction regions with a number of shells, each one serving a special purpose. The objective is to identify the particles and measure their energies or momenta and the direction of emission. Depending on the properties of different kinds of particles, various detection methods have to be applied. The different power needed to penetrate bulk matter provides a first and crude distinction of different kinds of particles. Some of them are absorbed in the innermost shell of the detectors, others can penetrate several metres of matter. This stopping power in matter provides only a crude but nevertheless very useful means to distinguish various kinds of particles.

According to the particles' penetrability of bulk matter, one can distinguish the following categories:

- *Photons and electrons*: Electrons carry an electric charge, whereas photons are electrically neutral. At high energies, however, they interact similarly with bulk matter; they both produce a cascade of charged particles (electrons, electron–positron pairs) and bremsstrahlung photons, which in turn are converted to particles. The probability of producing such a cascade and its geometrical length depend, of course, on the bulk material (heavy elements with high atomic number $Z$ are more efficient and give shorter cascades). Fortunately the lengths of electromagnetic cascades increase only slowly (i.e. logarithmically) with the energy of the incident particle and even at LEP energies they are at most about 50 cm long.[7] Therefore, electrons and photons are completely absorbed in about 50 cm of iron.

- *Hadrons*: Particles feeling the strong nuclear force such as protons, neutrons, pions, etc. produce in bulk matter a hadronic cascade consisting mainly of hadrons (with some photons originating mainly from the decay of neutral pions and some muons from the decay of charged pions). Surprisingly, in spite of the strong interactions these cascades are much longer than electromagnetic cascades (up to several metres), mainly owing to their content of neutrons, which are not readily absorbed (muons and photons also play a certain role).

- *Muons*: Their main interaction with matter is due to their electric charge. Penetrating material, they ionize atoms and in this way slowly lose energy. In contrast to electrons, they produce only little bremsstrahlung because of their greater mass and hence no electromagnetic cascades arise. As members of the lepton family they interact with matter also through the weak interaction, which, however, is so weak that it can be neglected.

- *Neutrinos* (or other exotic hypothetical particles): Because they carry no electric charge the only way for them to interact with matter is through the weak interaction. They can penetrate the earth or even escape from the interior of the sun and to detect them large specially designed detectors with many hundreds or thousands of tons of detecting material are necessary (e.g. to observe cosmic

---

[7] Electrically neutral pions mostly decay into two photons and their detection is therefore similar to that of photons.

neutrinos, but also neutrinos produced in accelerators). Such detectors cannot be integrated into the devices around the interaction regions of colliders. Hence, the existence of weakly interacting particles can only be inferred from 'missing energy' or 'missing momenta', exploiting the hermiticity of the detection device.

For the valid interpretation of events additional information has to be gathered. Tracks of particles must be observed to determine their directions, magnetic fields can be used to measure the momenta, special detector elements can provide information on the particle speeds or the arrival times of particles at particular places in the detector can be measured. The need to obtain such and other information dictates the basic patterns of the detectors. The different shells of a typical detector have different objectives, e.g. to determine the tracks of particles, their energy or momenta, and to identify them. Depending on the emphasis given to the diverse tasks, the arrangement of different shells will differ for various detectors.

A typical detector will consist of the following components, starting at the centre and going away from the collision region:

- *Vertex detector*: Some extremely short lived particles produced in the collision might travel only a few microns, even when travelling almost at the speed of light. After their short life they decay and they may be identified by the secondary decay particles. Such decays produce a vertex where the tracks of several particles meet. To detect and measure such vertices, a vertex detector has to be installed very close to the beam pipe and it must be able to measure tracks with extremely high precision.
- *Tracking chamber*: These devices can fulfil a double purpose – determine the direction of a particle and combined with a magnetic field measure its momentum. Several techniques have been developed to 'see' the tracks. The momenta of charged particles can be measured with high precision by their deflection in a magnetic field. Spatial resolutions of a few microns and high magnetic fields (e.g. by using superconducting coils) are necessary for this purpose. Sometimes these tracking chambers are divided into inner ones (close to the beam pipe) and outer ones (outer shell of the detector) to obtain a larger lever for the momentum measurement. These tracking chambers have replaced the previously used 'bubble chambers', which produced beautiful pictures of events but which had to be scanned by hundreds of people[8] to transform them into digitized format for the quantitative analysis. Track chambers, however, provide the tracks immediately in digitized format and, in addition, provide timing information. This allows preselection of events by a 'trigger' (see Sect. 8.4)

---

[8] These 'armies' of 'scanning girls' have disappeared, to the regret of many young physicists since they provided a nice opportunity to find a wife.

- *Electromagnetic calorimeters*[9]: The electromagnetic cascades produced by photons or electrons can be contained fully in some crystals (containing heavy atoms such as lead or bismuth), where they produce light flashes. The intensity of the flashes is proportional to the total energy deposited in the crystals and in this way the energy of the incident high-energy photons or electrons can be determined. For an absolute measurement of the energy, these calorimeters have to be calibrated in test beams.

- *Hadron calorimeters*: The hadronic cascades are too big to be captured totally in a homogenous detector (plastic scintillator or crystal), which would have to be several metres thick. A way out is to use sandwiches of thick slices of absorbing material interspersed by detecting layers (e.g. scintillators or devices measuring the ionization). If they are properly designed (and indeed an enormous amount of effort has gone into optimizing designs), the total scintillation or ionization output is proportional to the total energy of the incident hadron. Again, calibration is necessary for the measurement of the energy. In this case the term 'calorimeter' is even more inappropriate. When we used such a sandwich detector for the first time [4], I called it a 'sampling total absorption counter' (STAC), but this acronym did not become popular and 'calorimeter' became the standard name.

- *Cherenkov counters*: If a particle with a velocity close to the velocity of light traverses a dielectric material (e.g. water or glass) it is possible that the particle velocity is greater than that of light in the medium. In such a case the particle emits a shockwave of light, a type of radiation predicted by the Russian physicist Cherenkov.[10] The mechanism is similar to the production of the ultrasonic bang by fast airplanes or the bow wave of a ship. The magic blue glow which is known from pictures of nuclear reactors is also due to Cherenkov radiation.

  The simplest type of Cherenkov counter is the threshold counter, which provides an answer as to whether the velocity of a charged particle is lower or higher than a certain value by looking at whether the particle does or does not emit light. The most advanced type of Cherenkov detector is the ring imaging Cherenkov (RICH) detector, developed in the 1980s. In a RICH detector, the cone of the

---

[9] Particle physicists are not very clever in choosing their 'jargon'. The term 'calorimeter' is completely misleading. In chemistry a calorimeter is used to measure the total amount of energy released by an increase of temperature. In particle physics the total energy is measured by the output of scintillating light and no measurement of temperature is involved. Once a German minister of research, a chemist by training, visited CERN. When a CERN staff member explained a particle calorimeter to him in too detailed technical terms, the minister of research asked after some 10 min: "But I do not see any thermometers!"

[10] The velocity of light in the medium is reduced by the factor $1/n$, where $n$ is the refractive index of the medium, e.g. $n = 1.3$ for water. The opening angle $\theta$ (relative to the trajectory of the particle) of the light cone is given by $1/\cos\theta = n(v/c)$, where $v$ the velocity of the particle and $c$ the velocity of light in a vacuum. If $v < c/n$, no light is emitted. From the measurement of $\theta$ the velocity $v$ can be obtained. The momentum $p$ of a particle depends on its mass $m$ and velocity $v$ in the following (relativistic) way: $p = vm/[1 - (v/c)^2]^{1/2}$. If $p$ has been determined from the deflection in a magnetic field and $v$ from the Cherenkov angle, the mass can be calculated and the particle is identified.

Cherenkov light is intersected by a planar photon detector, where a ring of light is obtained, whose radius is a measure of the Cherenkov emission angle.

These are the most common types of detector components. Of course, they cannot all be optimized at the same time. The distinction between various experiments lies therefore in the different techniques utilized and the geometrical arrangement of the various components, thereby emphasizing different physics interests.

Finally, let me come back to the question of why the detectors have to be enormous. As already mentioned, the lengths of electromagnetic and hadronic cascades fortunately increase only slowly (logarithmically) with the energy of the incident particles. Even then the thickness of electromagnetic calorimeters has to be of the order of 0.5 m and that of hadronic calorimeters more than 1 m. In addition, space for the vertex detector, the tracking chamber and the magnetic coils is needed. Adding these dimensions together gives a 'radius' of several metres. There is, however, an additional important issue. The precision in measuring the momenta of particles is proportional to the product of the strength of the magnetic field and the length of the observed trajectory. Since the field strengths which can be produced over the large volume of a detector are limited (with superconducting coils several teslas have been obtained), one has to increase the track lengths as much as possible. As a result, the detector size scales linearly with the momentum (which at high energies is practically proportional to the energy) of the particles. These are the main reasons why detectors have to be so large for increasing energies.

In many fields miniaturization and nanotechnolgies have led to enormous reductions in size. Could particle physics not benefit from such developments? Miniaturization helps since it allows higher precisions in vertex or track chambers and this possibility has been largely exploited. The required accuracy in momentum measurements can then be achieved with shorter tracks. However, the basic laws of the interaction of particles with matter cannot be circumvented and large detectors are unavoidable.

## 7.2 The Four LEP Detectors

In this section the main characteristics of the four LEP detectors are briefly summarized. All four contain most of the various detection elements described in the previous section. Only those components which are specific for a detector will be mentioned. A feeling for the enormous sizes of the detectors can be obtained by comparing their sizes with those of the people standing next to them.

*ALEPH* (Apparatus for LEP Physics; Fig. 7.2) was a very ambitious project aiming to be a general-purpose detector able to cope with all physics at LEP. Excellent performance and reliability were the keywords. The central chamber, known as a time projection chamber (TPC), presented a new and risky technology to measure particle tracks because of its size. It consisted of a cylinder 4.4 m long and 3.6 m in diameter filled with a gas mixture of argon and methane. The particles traversing the TPC produce electrons along their tracks by ionizing the gas atoms and these

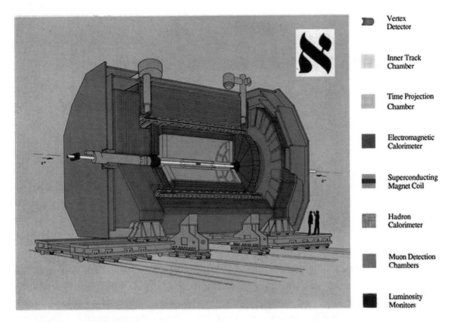

Vertex
Detector

Inner Track
Chamber

Time Projection
Chamber

Electromagnetic
Calorimeter

Superconducting
Magnet Coil

Hadron
Calorimeter

Muon Detection
Chambers

Luminosity
Monitors

**Fig. 7.2** The ALEPH detector

are dragged by an electric field (20 kV/m in a volume of 42 m$^3$) making them drift towards end plates of nearly of 10 m$^2$. They are segmented into individual pads with about 50,000 electronic channels. Measuring the drift time gives the coordinate of the track along the cylinder axis and the pad where the electrons hit the end plate gives the other coordinates. In this way the track can be reconstructed in three dimensions. A huge superconducting coil (length 6.4 m, diameter 5.3 m) producing a highly uniform magnetic field of 1.5 T was built by the CEA laboratory at Saclay near Paris. This magnetic field together with the bent tracks in the TPC give the momenta of charged particles.

   *DELPHI* (Detector with Lepton, Photon and Hadron Identification; Fig. 7.3) was conceived to put emphasis on the identification of hadrons and to determine the characteristics of leptons and photons. For this purpose RICH counters (using two different media) were installed which at the time of approval had never been employed before on such a large scale. At the heart of the detector a TPC was used for the detection of particle tracks (although smaller than the TPC of ALEPH). These detectors were mounted inside a superconducting solenoid with a diameter of 6.2 m and a length of 7.4 m producing a magnetic field of 1.2 T, making it the biggest superconducting solenoid in particle physics at that time. Its construction had been entrusted to the Rutherford Appleton Laboratory (RAL) in the UK. In contrast to the other detectors, in DELPHI the hadron calorimeter and the muon tracker were outside the coil.

**Fig. 7.3** DELPHI

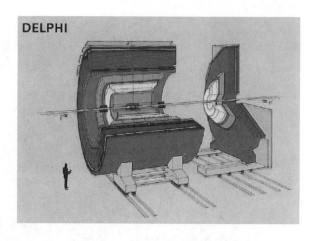

*L3* (third letter of intent; Fig. 7.4) was the largest of the four LEP detectors and was very special in identifying and measuring the momenta of photons and leptons. For that purpose a large electromagnetic calorimeter consisting of 12,000 bismuth–germanium oxide (BGO) crystals was created, a remarkable feature of L3. BGO was produced from very pure germanium oxide (about 10 tons) from the Soviet Union, the raw salt then went to China, where at the Shanghai Institute of Ceramics the crystals were grown. The electromagnetic calorimeter was followed by a 400-ton hadron calorimeter containing precision-shaped plates of depleted uranium delivered by the Soviet Union. L3 differed from the other detectors because its magnet coil was near the perimeter of the detector, whereas in the other detectors the magnet coil was close to the central detector. Thus, a high tracking precision for muons was obtained. To achieve such precision a giant octagonal magnet[11] 14 m long and 15.8 m high had to be built; it weighed 8,500 tons – as much as the Eiffel Tower! The muon momenta were measured by a barrel system comprising very large wire chambers (each 5.5 m long and 2.2 m wide) providing a spatial precision of 30 μm in spite of their large size.

*OPAL* (Omni-Purpose Apparatus for LEP; Fig. 7.5) like ALEPH was conceived as a general-purpose detector but using more conventional techniques following the experience with the JADE detector at the PETRA collider at DESY, Hamburg. To detect the positions of charged particles in the central part of the detector it used a 'jet chamber' containing almost 4,000 wires stretched longitudinally in a large volume of gas. When a particle flies nearby a wire it produces an electric signal in the wire and the position of the wire gives two coordinates. The coordinate in the longitudinal direction has to be determined by special 'z-chambers'. According to the OPAL concept to be a 'safe' conventional detector, a warm magnetic coil (magnetic field 0.4 T) to measure the momenta was used. However, OPAL had been

---

[11] This magnet is presently being used for the ALICE experiment at the LHC.

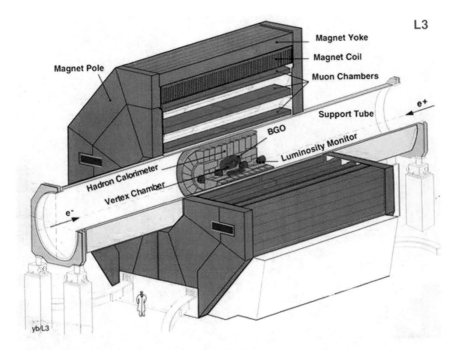

**Fig. 7.4** The L3 detector. *BGO* bismuth–germanium oxide

designed in such a way that this coil could easily be replaced by a superconducting coil producing higher magnetic fields, but this possibility was never used. To measure the energy of electrons and photons, about 12,000 blocks of lead glass (a well-proven material for the detection of photons) were used. The hadron calorime-

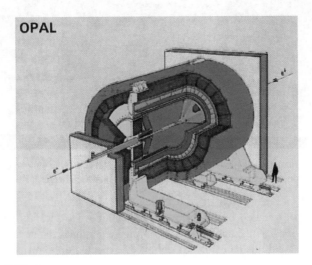

**Fig. 7.5** The OPAL detector

**Fig. 7.6** The front end of ALEPH. This photo demonstrates also the aesthetic aspects of technical equipment. In front of the detector is Nobel Prize laureate Jack Steinberger

ter consisted of a sandwich of iron plates and wire chambers in-between. The detector was surrounded by 10-m-long drift chambers to pick up and locate the penetrating muons. With such more conventional techniques the technological risks seemed limited and this detector was definitely expected to be ready at the turn on of LEP.

A few pictures of the real detectors might give an impression of their complexity. Figure 7.6 shows the front end of ALEPH. The superconducting coil of DELPHI was so large that it could not be transported to CERN using motor highways because of the limited height of bridges. It had to come on small roads through the Jura Mountains. To squeeze it through some small villages was a problem and at one place a corner of a house had to be removed (Fig. 7.7). Figure 7.8 shows the magnet of L3 with some of the collaborators assembled around it. Figure 7.9 shows one half of the impressive barrel of the lead glass crystals of OPAL used to record hard photons.

Figure 7.10 shows a typical installation of a detector in the underground hall and the excess shafts. The components, after having been brought down through the main shaft (experimental access shaft), could be assembled in the experimental cavern and when completed the detector could be rolled into the final position into the LEP beams. This scheme does not apply to L3, whose magnet was to large and heavy to move.

When approving the detectors, we thought to strike a good balance between well-proven techniques and new ones which were still under development. To design the detectors, to build and to install them was a major challenge involving many universities, national institutes and CERN. Thanks to the enormous enthusiasm, stubborn engagement and competence of scientists, engineers and technicians, to

**Fig. 7.7** The largest superconducting solenoid ever produced and destined for DELPHI had to be transported to CERN on roads without bridges but difficulties were still encountered in narrow passages

**Fig. 7.8** The magnet of L3 surrounded by some of the collaborators

my surprise all the detectors were ready when LEP started operation and they immediately started to produce exiting physics results.

## 7.3 Data Acquisition and Evaluation

The particle bunches circulating around LEP with almost the velocity of light were crossing about 50,000 times per second (every 20 μs). However, most of these crossings resulted in events which were of known nature and therefore of no interest (e.g. elastic scattering). In general, one could expect one 'good' event per second. Every event could involve tens of megabits of information and hence it would have been practically impossible to record all events and store all the data. Therefore, fast decisions had to be taken on whether an event was interesting enough to be recorded or whether it should be discarded immediately. Indeed, a whole chain of successively more detailed decisions had to be taken by fast electronic circuits and these kinds of selections are called 'triggers'. At the first level, one must decide within 20 μs, i.e. before the next beam collision, if an event is potentially interesting. This decision is based on fast time coincidences between various detector elements. At the second level, time-sliced readouts of detector elements are used for a more rigid selection within about 100 μs. At this stage the full data for an event are more or less available and hundreds of milliseconds has to be spent transferring the data to a powerful computer. After the first preliminary on-line analysis, the data are put on magnetic tapes at a rate of about one or two events per second. Such on-line

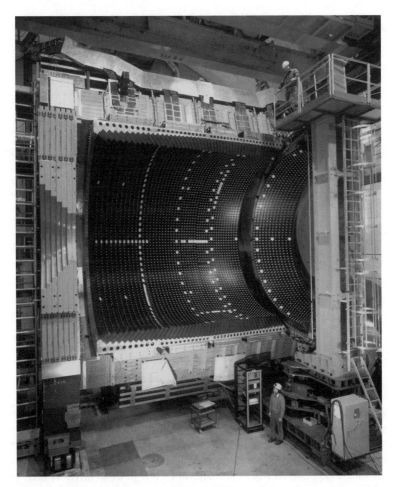

**Fig. 7.9** The lead glass calorimeter (one half) of OPAL consisting of several thousand crystals used to detect photons

data handling became possible thanks to the development of powerful computers[12] which were installed in the CERN computer centre. The final off-line data analysis of these raw data could then be done at the home institutions. For this purpose it had to be possible to transfer the data from CERN to the outside institutes and they had to have access to the service programmes available at CERN. This need was the main motivation to develop the World Wide Web.

Finally it should be mentioned that for the measurement of absolute event rates (absolute cross-sections) the precise number of collisions has to be known. This is given by the luminosity (see Chap. 6). The beam conditions might be slightly

---

[12] More details can be found in [5].

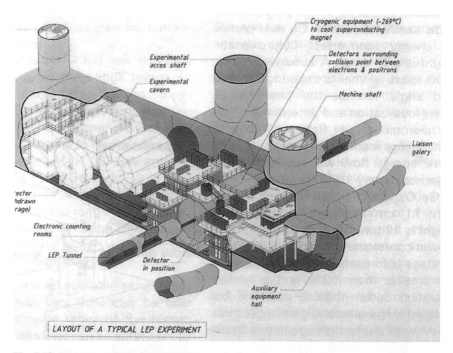

**Fig. 7.10** Typical installation of a LEP detector in the underground hall

different for each of the detectors and they change over time with the tuning of the beams. Hence, each experiment had to monitor the integrated luminosity permanently, which gave a measure of the performance of the LEP machine and of the particular experiment, including the solid angle covered by the detector, the trigger efficiency and many other parameters. Therefore, it is not surprising that the careful analysis of the data sometimes took months or even years to arrive at final precise results.

## 7.4 Organization and Management of the Collaborations

The LEP detectors presented an entirely new problem for CERN. Because of the limitations of the CERN budget, only a relatively small contribution to their financing could be provided by CERN. For previous detectors or experiments the outside contributions were in general small, with the result that CERN could keep full control over the construction and operation of facilities.[13] However, for the LEP

---

[13] There were a few exceptions. For example, the Big European Bubble Chamber (BEBC) was mainly financed by France and Germany; however, the arrangements were such that CERN could keep control.

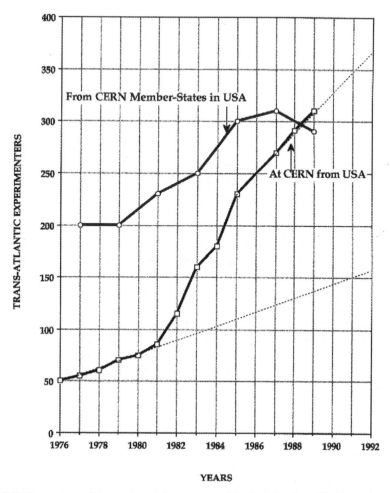

**Fig. 7.11** Comparison of the number of elementary particle physicists from CERN member states working in the USA with the number of US physicists participating in CERN experiments

experiments the situation was completely different. More than two thirds of the components were to be provided by outside institutions. The total cost was estimated to amount to about CHF 340 million (47% from member states and 38% from non-member states) and CERN could contribute only CHF 50 million (15%). The most severe hazard was that CERN had no reserves to help in the case of technical or financial difficulties. As a result, CERN had little control over the construction of the detectors, each being a big project in itself. A new mentality had to be introduced. The collaborations had to learn that the participating institutions had a common responsibility.

To put this on a realistic basis, a special management structure had to be created for each collaboration.[14] It was governed by a steering committee composed of representatives of each participating institution. The steering committee appointed a spokesperson, charged with representing and managing the collaboration, and a technical coordinator from CERN responsible for the overall technical management. In addition, I insisted that a Collaboration Finance Review Committee be established for each of the collaborations. Since scientists are sometimes too optimistic, the members of these committees had to be nominated by the funding organizations in the various countries. Contributions by the collaborating institutions to the detectors were primarily made 'in kind', meaning that fabricated components of the detectors were delivered. However, some items had to be financed commonly by the whole collaboration and for this purpose 'common funds' were set up. They were also used as a reserve for those cases in which one partner could not provide the promised contributions. Multilateral agreements with the participating institutions provided in the end a proper legal basis for all these arrangements. I called this way of organizing an international project the 'LEP model', and this was extended recently to the LHC to provide not only parts of the detectors but also of the machine.[15]

It might be justified to say that with LEP and its international exploitation, CERN has become a laboratory serving the worldwide community of elementary particle physicists although formally still a European organization. In particular, LEP helped to reverse the brain drain from Europe to the USA. During the preparation of LEP experiments, the number of US scientists at CERN increased rapidly and a crossover with Europeans working in the USA happened when LEP came into operation (Fig. 7.11).

## References

1. Bourquin M (ed) (1981) ECFA report on the general meeting on LEP, Villars-sur-Ollon, 1–7 June 1981, ECFA 81/54
2. Robinson AL (1982) CERN gives nod to four LEP detectors. Science 217:722
3. Fabjan C, Schopper H (2009) Handbook of particle physics, vol 2. Springer, Heidelberg
4. Engler J, Flauger W, Gibbard B, Mönnig F, Runge K, Schopper H (1973) A total absorption spectrometer for energy measurements of high-energy particles. Nucl Instrum Methods 106:189
5. Butterworth I (1986) Supercomputers make their mark. CERN Cour 26:6

---

[14] CERN Finance Committee FC/2801, 4 December 1984

[15] Such a model was first used to a large extent at DESY for the construction of HERA, for which half of the bending magnets were provided by Italy.

# Chapter 8
# What Have We Learned from LEP? – Physics Results

Whenever a new facility for elementary particle physics is designed and constructed, it is a step into the unknown. It is the real purpose of such machines to venture into unexplored territory – that is why it makes working with them so exiting. Of course, there is always some guidance from theory. An essential element of progress in science concerns the falsification of theories. The imagination of theorists is overwhelming and an important role of experiments lies in disproving most of the theories en vogue and singling out the right ones. On the other hand, experiments can confirm the predictions of a theory. A theory is usually only accepted as 'better' than an existing theory if it describes all known experimental facts and makes, in addition, predictions for new phenomena which can be tested experimentally. Between two competing theories, the one is preferred which starts from fewer assumptions. This is called Ockham's razor after the philosopher Wilhelm von Ockham (1285–1349). Mathematics is used as the language to express the physical laws unifying the incredibly rich variety of experimental facts.[1] This kind of progress implies a continuous interaction between theory and experiment and involves hard work over many years.

Of course, sometimes it happens that something unexpected is found by experiments which makes the headlines, sometimes even in popular newspapers. But what does 'unexpected' mean? 'Unexpected' in physics means unforeseen relative to an accepted theory. But even the most surprising discoveries do not destroy the established knowledge – there are no real revolutions in physics even if the media like to present it that way. What rather happens is the discovery that an 'old' theory is not universally valid. It remains valid in a restricted domain. Einstein's special theory of relativity, for example, showed that the classical Newtonian theory can be used only for phenomena at velocities that are low compared with the velocity of light. For high velocities Einstein's theory has to be used and contains the classical theory as a

---

[1] A science only becomes a 'quantitative' science when its results can be formulated in mathematical terms. Thus, chemistry became a 'real' science only when chemical processes could be described by quantum mechanics. Unfortunately the need for mathematics as an essential tool prevents many people from enjoying the magnificent world of science because of their aversion to mathematics.

H. Schopper, *LEP – The Lord of the Collider Rings at CERN 1980–2000*,
DOI 10.1007/978-3-540-89301-1_8, © Springer-Verlag Berlin Heidelberg 2009

special case. Similarly quantum mechanics has to replace classical physics when dimensions of the size of atoms come into play, whereas at large dimensions classical physics remains valid.

When LEP was proposed the expectations went in both of these directions. LEP 1, with beam energies of 50 GeV, was expected to verify with high precision the standard model of elementary particle physics, which before LEP was still standing on weak ground and small deviations could point to new phenomena beyond the standard model. The second direction of research concerned the search for new particles predicted by some theories. In particular, there was the hope that LEP 2, with beam energies about twice as high as those of LEP 1, would produce some of the hypothetical particles predicted, for example, by the so-called supersymmetry theories or string models. Even one of the essential elements of the standard model was still missing, the Higgs particle. The difficulty is that all the current theories cannot make any predictions for the masses of these unknown particles and therefore the experimenters do not know at what energies to look for them.

In summary, one might state that LEP surpassed all expectations as far as the verification of the standard model is concerned. LEP converted high-energy physics from a '10% science' into a field of very high precision. In technical terms it showed with a precision of better than fractions of a percent that the standard model is a 'renormalizable field theory', which led to the award of the Nobel Prize to the two theoreticians Martinus Veltman and Gerardus 't Hooft.[2] On the other hand, no spectacular discovery could be made at LEP, be it completely unexpected or supporting some of the theoretical predictions made by theories going beyond the standard model. It turned out that the collision energies producible at LEP were too low for these new phenomena. However, some of those theories could be excluded and a few important hints in which directions to look in the future could be obtained. The legacy of LEP is nevertheless quite remarkable. Twelve years of hard work at LEP produced a coherent set of very accurate and self-consistent results going far beyond the original expectations. The LEP measurements, more precise by orders of magnitude that what was previously available, put the standard model on a safe basis and provided guidelines for future investigations. This is particularly true for the energy range reachable by the new machine in the LEP tunnel, the LHC.

---

[2] On this occasion a problem with the Nobel Prize has again become obvious. The Nobel Committee adheres to the tradition (although not fixed by Nobel's testament) to award the Nobel Prize to no more than three scientists. Since in many fields, not only in elementary particle physics, experiments are often carried out by large collaborations with many leading scientists, it happens more and more often that theorists, who can still work alone, are awarded the Nobel Prize, but it is certainly well deserved. However, this practice might in the long run work against the smooth running of large collaborations. My effort to convince the Nobel Committee to attribute the Nobel Prize to collaborations (as is done for the Nobel Prize for Peace) although justified met with no success. The argument was that the public wants to see heroes. Certainly to 'sell' science to the public is easier when it can be personalized. If the Higgs particle is detected, the theoretician Peter Higgs will be awarded the Nobel Prize but the experimenters who discover the Higgs particle will not be awarded it.

It is not the purpose of this book to give a detailed report of the enormous quantity and quality of the LEP results. Only some highlights[3] will be presented which will demonstrate how successful LEP was.

## 8.1 What Is the Standard Model?

Even among physicists discussions are going on concerning the different meanings of 'model' and 'theory'. A real theory is a theoretical framework starting from a few fundamental assumptions from which the results expected from experimental observations can be mathematically deduced. A theory is considered to be better if more observations can be reproduced by fewer assumptions. A model is also a mathematical framework with the main aim to reproduce specific experimental results. It is rather phenomenologically oriented, making many assumptions and introducing a number of free parameters which have to be determined from the experimental data. It can explain much of the known world, yet it is still work in progress. Usually a theory also needs some numerical input from observations, but only for a very limited number of parameters. With progress, a model can be converted into a real theory in which many assumptions and parameter values will come out automatically. Of course, the boundary between a theory and a model is somewhat fuzzy.

The standard model of particle physics has certainly evolved into more than a simple model. It provides a detailed mathematical framework encompassing practically all of the present experimental results concerning the subatomic particles and the forces between them. It is based on a concise set of principles and mathematical equations derived from them. The limit of human ingenuity but also perseverance and tedious work both in theory and experiments were necessary to achieve this success representing one of the most impressive steps in understanding nature. It is not the purpose of this book to review the standard model and in particular its mathematical framework.[4] Only a few basic concepts will be mentioned which are necessary to appreciate the results obtained by the LEP experiments. However, the standard model leaves open a number of fundamental questions and therefore cannot be the final theory, 'the theory of everything' as it is sometimes called. The consolidation of the standard model and the search for new phenomena outside it are the two most thrilling directions of research in particle physics.

## 8.2 Building Blocks of Matter

One of the fundamental questions concerns the building blocks of matter. After chemists had established the periodic system of elements, atoms were detected. However, contrary to the meaning of the Greek word (the 'indivisible'), atoms can

---

[3] A comprehensive survey is given in [1].

[4] For concise recent summaries, see [2, 3]. For a detailed review, see [4].

be split and they consist of a nucleus containing most of the mass and an electron cloud surrounding it. In the next step to smaller units of matter it was found that the atomic nucleus consists of protons and neutrons. Protons and neutrons together with the electrons, the carriers of the electric current, were considered to be the three final building blocks of matter until the first half of the last century (see Fig. 1.1). In the 1960s a whole 'zoo' of new 'elementary particles' was discovered, making the previously simple picture very complicated. Yet, three decades of intensive research have produced again a rather simple picture.

Our present conception of matter comprises two categories of building blocks. The proton and neutron are not elementary but consist of various kinds of quark. The other family of the building blocks of matter is made up of the leptons, whose best known member is the electron. The electron has heavier 'brothers', the muon and the tau particle. Each of these charged leptons has an electrically neutral partner, the various kinds of neutrino (Fig. 8.1). The 12 'matter particles' can be arranged in a beautiful scheme with many symmetries. The individual particles are characterized by the charges they carry and these determine the systematic character of this 'periodic system of elementary particles' which replaces in a way the well-known periodic system of elements as established by chemists. It is even much simpler.

There are three families of quarks and in a symmetric way three families of leptons each consisting of a doublet. The members of a doublet in the upper or lower row of the periodic system, respectively, carry electric charges which differ by one unit of the elementary electric charge. The quarks have 2/3 or −1/3 charges, whereas the leptons have 0 or −1 charges. The three columns in each of the families are distinguished by their 'flavours', a kind of weak charge (sometimes one speaks of different 'generations' in a family). The doublets are arranged in such a way that the lightest are on the left and their masses increase with each generation to the right. Often one refers to the individual particles by giving them names. For the quarks, up (u), down (d), charm (c), strange (s), top (t) and bottom (b), and for the leptons, electron (e), muon (μ) and tau (τ) and electron-neutrino ($v_e$), muon-neutrino ($v_\mu$) and tau-neutrino ($v_\tau$).

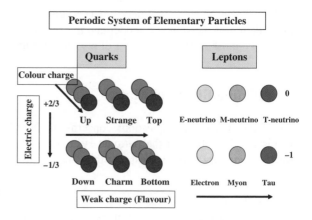

**Fig. 8.1** The periodic system of elementary particles

We still have to explain what distinguishes the family of quarks from the family of leptons. The quarks carry a kind of charge which is akin to the electric charge and is called 'colour'. This name is metaphoric and has nothing to do with the colours of our daily life. This term was chosen because in contrast to the electric charge, the colour charge appears with three different values, which are given colour names, for example blue, green and red. This complication opens a new way to create neutral or 'colourless' particles. In electricity a neutral state can be produced by a charge and its anticharge, combining minus and plus. In the same way a colourless state can be obtained by combining a red (or blue or green) with an antired (or antiblue or antigreen) charge. However, a colourless particle can also be achieved in another way, by combining all three colour charges. This is similar to producing the colour white by mixing the three basic colours and this is exactly why colour names were chosen to distinguish quark charges. There is another fundamental difference between electric and colour charges which might at first seem somewhat confusing. These three colour charges correspond to the only negative charge in electricity. The positive electric charge is not a second kind of charge but it is simply the anticharge of the negative charge. In a similar manner each of the three colour charges has its own anticolour charge.

Nature uses both possibilities to produce colourless particles. Mesons are particles consisting of a quark and an antiquark with opposite colour charges, resulting in a 'white' particle. The many possible combinations of quarks with antiquarks lead to the 'zoo' of many mesons (e.g. the pion and the K, D and B mesons). Nucleons, on the other hand, are particles consisting of three quarks carrying all three colour charges and in that way give 'white' particles. The most prominent representatives of this group are the proton (uud) and the neutron (udd). The common name for mesons and nucleons, i.e. for these 'white' particles, is 'hadrons'.[5] So far only 'white' particles have been found in nature and it seems that the existence of free coloured particles is forbidden. This implies that no free quarks can be observed and that their existence and properties can only be studied inside 'white' particles.[6]

It was discovered some time ago that besides matter also antimatter exists, which implies that each of the matter particles has a 'mirror' particle for which all charges have opposite signs.[7] For example, the antiparticle of the electron has a positive electric charge (hence its name 'positron'). When a particle meets its antiparticle, they can annihilate each other since charges with opposite sign neutralize each other. In such an annihilation process mass is converted completely into energy and this occurrence is, of course, the basis for an electron–positron collider like LEP (see

---

[5] The name 'hadron' comes from the Greek word for 'strong', indicating that the quarks inside these particles are bound together by the strong nuclear force.

[6] One of the exiting possibilities is to produce a quark–gluon plasma by the collision of heavy nuclei. Inside such a plasma quarks and gluons would move freely.

[7] The concept of antimatter became known to a large part of the population from Dan Brown's thriller *Angels and Demons* [5], where the antimatter activities at CERN play a major role. To find out the difference between fiction and reality see http://www.cern.ch.

Chap. 1). In the same way as a positive charge is the anticharge of negative electricity, one has to introduce anticolour charges as already mentioned.

From these concepts simple but far-reaching conclusions can be drawn. What we call 'negative' or 'positive' electric charge was an arbitrary historical decision. Our predecessors 200 years ago could have made the opposite choice. Since nature 'does not know' what choice was made, the laws of nature should be independent of this choice – there should be a symmetry between negative and positive charges. In other words, if all the negative charges in the universe were replaced by positive ones and vice versa nothing should change. The same argument applies to the colour charges and we expect a similar symmetry. Since matter and antimatter particles are distinguished by the opposite signs of their charges, there should also exist a symmetry between matter and antimatter. Indeed, in many laboratory experiments these symmetries were found and confirmed.

Because of the symmetry relating matter and antimatter, one would expect that at the beginning of the universe equal amounts of matter and antimatter were created. This probably happened, but when the matter and antimatter met each other most of it was annihilated again and converted into massless energy (e.g. light or neutrinos). It is one of the great mysteries why celestial objects (and with them also we humans) exist at all. Why have only galaxies made of matter been found by astronomers? Why have no antiworlds been detected in the universe? During the great annihilation at the beginning of the universe, apparently a tiny amount of surplus of matter was left over and this is the stuff of which we are made. Why such an asymmetry between matter and antimatter occurred is one of the great riddles waiting to be explained. Maybe particle physics can provide the key for its solution.[8]

## 8.3 The Forces of Nature

Various forces act between the 12 'matter particles' grouped into the two families, the quarks and leptons (and their antiparticles). The longest known is gravitation. It can be neglected when dealing with elementary particles since their masses are so small that gravitational attraction can be ignored with respect to the other forces.

In the nineteenth century the electric and magnetic forces were discovered and one of the big achievements was their unification into one force, the electromagnetic force – basic knowledge which became the foundation of all modern electrical technologies. In the twentieth century two new forces were found, the strong nuclear force keeping the protons and neutrons together in the atomic nucleus and the weak nuclear force giving rise to nuclear β-decay and being essential for energy production in the sun. Other forces may exist in nature which have not yet been discovered.

---

[8] One conjecture is that the small violation of the so-called PC symmetry of the weak nuclear force could be at the origin of this asymmetry.

The unification of gravitation with the other forces is one of the most fundamental unsolved problems. The main difficulty arises from the fact that the other three forces obey quantum mechanics, whereas gravitation does not. 'String theories' replacing pointlike objects such as quarks and leptons by fundamental extended objects, the 'strings', try to achieve this goal.

The sources of the forces are charges, which were introduced above: the electric charge for the electromagnetic force, the colour charges for the strong nuclear force and flavours for the weak nuclear force. The forces can be described by fields which obey certain laws of nature expressed in mathematical form by (e.g. differential) equations. For the electric field these are the famous Maxwell equations.

For a long time one wondered how these forces could act over large distances. In classical physics acoustic waves are propagated in air and electric fields were thought to be carried by the somewhat mystical ether which was abolished by the theory of relativity. But in empty space there is no medium to transmit the forces, an enigma in classical physics. In terms of modern quantum field theories the action of forces can be understood by the exchange of 'carrier particles' (Fig. 8.2) which are called 'field quanta'. The matter particles 'play ball' by emitting and capturing these carrier particles. One might visualize their actions as an exchange of balls between two boats. Catching and throwing the balls produces recoils which, seen from a distance, are interpreted as a repulsion. Each force field is produced by its specific charges and is transmitted by its carrier or binding particles, the field quanta.

The carrier of the electromagnetic force is the photon (predicted by Albert Einstein in 1905 on the basis of Max Planck's theory, but experimentally confirmed only around 1920). The photons (or the quanta of light) transport not only the energy from the sun to the earth through empty space, but they also act in all electromotors and generators by transferring the force. Photons have no electric charge and hence there are no forces between photons themselves.

The strong nuclear force is transmitted by gluons, which bind the quarks together and hence their name. There are eight different kinds of gluon and they are distinguished by different combinations of a colour and an anticolour charge.[9] Gluons can interact with the colour charges of quarks, but also between themselves since they carry colour charges. This gluon–gluon interaction is a special characteristic of the strong nuclear interaction resulting, for example, in the 'confinement' of quarks and gluons, i.e. they cannot be observed as free particles. The existence of the gluon was confirmed by experiments at the PETRA collider at DESY in 1976.

The carriers of the weak force are the electrically neutral Z and the electrically charged W particles. They carry weak charges and hence interact with the weak charges of quarks and leptons, but they also interact between themselves. In contrast to the photon and the gluons, which are massless, the Z and W particles are very

---

[9] There are nine combinations of the three colour and three anticolour charges. According to group theory, they can be split into an octet, the eight gluons, and a singulet. The singulet would be colourless and such a gluon could appear as a free particle, which has not been observed.

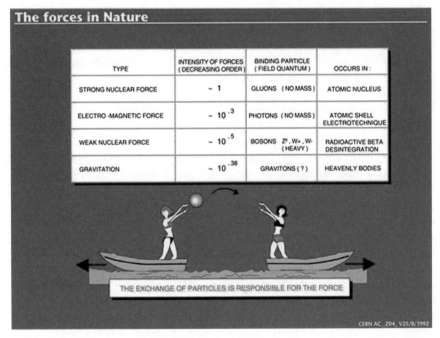

## The forces in Nature

| TYPE | INTENSITY OF FORCES (DECREASING ORDER) | BINDING PARTICLE (FIELD QUANTUM) | OCCURS IN: |
|---|---|---|---|
| STRONG NUCLEAR FORCE | $\sim 1$ | GLUONS (NO MASS) | ATOMIC NUCLEUS |
| ELECTRO-MAGNETIC FORCE | $\sim 10^{-3}$ | PHOTONS (NO MASS) | ATOMIC SHELL ELECTROTECHNIQUE |
| WEAK NUCLEAR FORCE | $\sim 10^{-5}$ | BOSONS $Z^0$, W+, W- (HEAVY) | RADIOACTIVE BETA DESINTEGRATION |
| GRAVITATION | $\sim 10^{-38}$ | GRAVITONS (?) | HEAVENLY BODIES |

THE EXCHANGE OF PARTICLES IS RESPONSIBLE FOR THE FORCE

CERN AC_Z04_V25/8/1992

**Fig. 8.2** The forces in nature and their carrier particles. Forces are exerted by the exchange of field quanta. Their strength is determined by the respective coupling constants

heavy and to produce them very high energies are necessary. A small number[10] of them were produced at the CERN proton–antiproton collider in 1983, but their detailed properties could be studied only at LEP.

The charges determine the strengths of the forces between matter particles and therefore the charges are sometimes called 'coupling constants'.[11] This is, however, a misleading term since, as we shall see later, the coupling constants change with the interaction energy. Sometimes the question has been raised of whether the strength of the forces (or 'interactions' as physicists prefer to call them) has changed with the age of the universe. This can be investigated by the observation of far-away stars which are much younger than our galaxy. No variation of the coupling constants has been observed.

Another fundamental quantity characterizing particles is their spin. The spin of a particle can be imagined as indicating its rotation around a fixed axis. However, for elementary particles it is a purely quantum mechanical quantity with no classical

---

[10] It was only half a dozen, but sufficient for their discovery and leading to the award of the Nobel Prize to Carlo Rubbia and Simon van de Meer, both at CERN.

[11] For their easier use and to express their fundamental character, the charges are often multiplied by other fundamental constants to make them independent of measuring units (dimensionless) (see Sect. 8.6.1). This is not possible for gravitation since it does not obey quantum mechanics.

analogue. Spin is given in units of the Planck constant $h/2\pi$ and in this unit spin can only assume integer and half-integer values. All matter particles have spin 1/2, whereas the carriers of the forces, the field quanta, have spin 1 (with the exception of the still hypothetical quantum of gravitation, which should have spin 2). This is a fundamental difference between matter and forces[12] and in the theories requiring 'supersymmetry' one tries to abolish this difference by establishing full symmetry between matter and interactions.

## 8.4 Symmetries – the New Paradigms

A development which makes particle physics particularly interesting, even from a philosophical point of view, is the fact that during the past decades the paradigms used to understand nature have changed. After Democritos introduced the idea of the atom and Newton spoke about 'infinitely hard spheres' progress in understanding nature was linked to the discovery of ever smaller building blocks of matter (see Fig. 1.1). The main paradigm was based on two elements: 'eternal' building blocks of nature and forces which act between them. Thanks to the forces, the building blocks can be put together in different combinations, thus creating the permanent changes, including birth and death.

The last century brought a complete new basis for the laws of nature – symmetries. I consider these to be more exciting and fundamental than the discovery of a new particle. Unfortunately symmetries are more abstract and hence more difficult to explain to a layperson than concrete building blocks – most people prefer concreteness to abstraction. In an oversimplified way, one may characterize this fundamental change of paradigms by stating that we move away from the atoms of Democritos to abstract 'ideas' of Plato. Symmetries are considered to be the ultimate principles to understand the microcosm and in a perfect theory the various kinds of matter particles and their properties as well as the forces and their characteristics should be deductible from first principles.

What are symmetries in physics? The symmetry we know best from daily life is mirror symmetry. A mirror symmetry was mentioned earlier for the electric and other charges. If we replaced all the charges in the universe by their anticharges, neither the earth nor the laws governing it should change. Another example is a sphere, the most symmetrical object known. If we turn a sphere around an arbitrary axis nothing changes or, what has the same effect, we could leave the sphere unchanged but look at it from different angles. A cylinder is less symmetric than a sphere, it can be rotated only around one preferred axis.

---

[12] Particles with half-integer spin obey Fermi statistics, implying that quantum states can be occupied by only one particle. Particles with integer spin obey Bose statistics, with the consequence that any number of particles can occupy a quantum state. Fermions can only be produced as particle–antiparticle pairs, whereas bosons can be produced singly.

If we talk about a symmetry principle in physics it means that if a special operation is performed the laws of nature do not change. For some symmetry operations we would think that they should be valid a priori. One example has been given already. Since our choice of what we call a negative or a positive electric charge is arbitrary, a mirror operation with respect to the sign of the charge should leave the laws of nature invariant. In addition to mirror operations, which correspond to a flip-flop, there are other 'continuous' changes which we can introduce. To perform measurements we have to choose a coordinate system relative to which we determine distances. We would think that the laws of nature should not depend on the choice of the coordinate system; hence, these laws should be independent ('invariant') of a shift or a rotation of the system. It has been known for a long time (but not appreciated so much, and has never been taught in schools!) that the invariance with respect to translations and rotations has two fundamental consequences in physics – the conservation of linear momentum and of angular momentum. Nature does not care at what particular moment we start our clock for an observation; therefore, the laws of nature should also be invariant against a translation in time, and this symmetry immediately has another fundamental consequence, the conservation of energy!

These symmetries which dominate classical physics are easy to visualize (they are *anschaulich*) and should be valid from an a priori point of view. In quantum mechanics other more abstract symmetries can be considered. Whether they are valid or not cannot be said from the beginning; hence, theorists make assumptions about the fundamental symmetries and deduce from them a particular theory which then has to be tested by experiments. Here we see the very productive cooperation between theory and experiment.

## 8.5  The Symmetries of the Standard Model

For a long time Einstein's theory of special relativity (valid for phenomena at high speeds) and quantum mechanics (valid in the atomic realm) existed independently side by side. The development of so-called renormalizable quantum field theories offered a frame to bring the two theories together. Both are necessary in particle physics since the laws of quantum mechanics have to be followed for subatomic events and special relativity has to be applied because of the large energies and high velocities involved. Apart from the classical 'symmetries' considered in Sect. 8.4, a new abstract principle is introduced: local 'gauge' invariance. One of the counterintuitive properties of quantum mechanics is that particles are also described by wave fields (particle–wave duality[13]). 'Gauge invariance' means that the fields associated with the various particles are invariant under transformations changing some abstract internal parameters at a particular point in space and time (hence the name

---

[13] Vice versa the force fields are quantized, they contain field quanta. These are the force carriers introduced in Sect. 8.3.

'local').[14] These 'gauge' theories start from the assumption that the fundamental particles and charges are pointlike. This leads to some mathematical anomalies ('divergencies') which can be removed by clever mathematical 'tricks'[15] which are called 'renormalization'. The mathematical framework of these theories is rather complicated and is beyond the scope of this book.

Within this general framework many different theories are possible and they are distinguished by the special gauge symmetry from which they start. At present no fundamental principles are known to indicate which particular gauge group should be preferred and the only possibility is to guess and find out which one reproduces the experimental results in the best (with high precision) and simplest (starting from as few assumptions as possible) way. This led to the development of the standard model, which is based on the symmetry group $SU(3) \times SU(2) \times U(1)$. Here $SU(3)$ is the 'colour' group of the theory of strong interactions with three colour charges and $SU(2) \times U(1)$ describes the electroweak interactions. $U(1)$ is associated with the electromagnetic field with the electric charge $Q$. The eight massless gluons are associated with $SU(3)$, while for $SU(2) \times U(1)$ there are four gauge bosons $W^+$, $W^-$, $Z^0$ and $\gamma$. Of these, only the photon $\gamma$ is massless, whereas the three bosons associated with the weak interaction have large masses.

To explain these empirical facts an additional concept had to be introduced – 'spontaneous symmetry breaking'. In principle, this phenomenon is well known from the magnetization of iron. A piece of soft iron is not usually magnetized. Below a certain temperature (the Curie temperature), however, the iron becomes spontaneously magnetized even without an external magnetic field. In the unmagnetized state the little elementary magnets inside the iron point with equal probability in all directions and the overall magnetization is cancelled. No direction in space is preferred and the iron is in a perfectly symmetric state. After the spontaneous magnetization, suddenly one direction in space is favoured, and the symmetry is broken. A similar phenomenon happens when putting a pencil on its tip. For a moment it remains in the vertical position, but it will rapidly fall down in an arbitrary direction. Again, the initially symmetric state is spontaneously broken. In more technical terms it implies that the interaction still obeys a certain symmetry (the interaction potential is symmetric) but for the observable state it 'chooses' a particular possibility from an infinite number of equally probable ones.

In the frame of the standard model such a spontaneously broken symmetry mechanism is nothing other than the famous Higgs mechanism introduced by Peter Higgs in the 1960s. It requires the existence of a new particle, the Higgs particle. The Higgs mechanism does not only give a heavy mass to the Higgs particle itself, but also

---

[14] The gauge invariance is well known for the electromagnetic field. One can measure electric voltages, which are differences between electric potentials. The zero level of the potential can be chosen arbitrarily, it can be 'gauged', without changing the measured voltage. This apparently trivial invariance has the result that there is a conservation law for the electric charge.

[15] These 'tricks', however, are so clever that they gave rise to various awards of the Nobel Prize when they were applied to the electromagnetic force for the first time and later to the electroweak force.

gives masses to all other particles, including the W and Z bosons, and also the matter particles, the quarks. However, the present theories cannot make any predictions about the mass of the Higgs particle or any other masses.

One of the great successes of the standard model is the partial unification of the electromagnetic and the weak force by the group $SU(2) \times U(1)$ and today one speaks of the 'electroweak force'. The unification, however, is not complete. For example, the relation between the strength of the various forces should be predicted by a full theory, but the standard model cannot achieve this. The strong interaction described by $SU(3)$ is somewhat arbitrarily 'glued' to the group $SU(2) \times U(1)$ associated with the electroweak interaction. Some other shortcomings of the standard model will be discussed later. However, the great success of the standard model is given by the fact that so far it has reproduced all experimental observations with great precision. In particular it could predict the existence of the force carriers W and Z. The experimental verification of the standard model concerns, therefore, above all a detailed investigation of these particles. The extremely precise LEP experiments achieved such a verification and have put the standard model on a firm basis. In addition, many results have been obtained for the strong force. Not all of these results can be described here and only some of the main achievements can be mentioned.

## 8.6  The Z Factory – Results from LEP 1

During its first phase of operation from 1990 to 1996, LEP operated with beam energies around 50 GeV (LEP 1), with the main objective to produce large quantities of Z particles, one of the carriers of the weak force. The production probability is largest if the sum[16] of the energies $E_{total}$ of the colliding particles precisely matches the mass of the Z particle according to the famous Einstein equation $E_{total} = m_Z c^2$, where $m_Z$ is the mass of the Z particle and $c$ is the velocity of light. If this condition is met, one talks of 'resonance production'.

Once a Z particle has been produced, it decays after an extremely short time. Since conservation laws for the energy and momentum and for the electric and other charges must be obeyed, it can only decay into a pair consisting of a fermion and an antifermion, which are emitted in exactly opposite directions. This two step process can be described by $e^+e^- \rightarrow Z \rightarrow e^+e^-$ or $\mu^+\mu^-$ or $\tau^+\tau^-$ or quark and antiquark.

Around 20 million Z particles were observed with LEP, elevating the Z particle to one of the best studied objects in particle physics and allowing the determination of a number of fundamental parameters of the weak force. A $Z \rightarrow \mu^+\mu^-$ event is shown in Fig. 8.3. As the figure shows, such events are extremely clean, showing besides the two particles no other tracks, i.e. no background, one of the big advantages of an electron–positron collider.

---

[16] Strictly speaking the total energy in the centre of mass is the relevant quantity and is equal to the sum of the energies of electrons and positrons provided they have the same energy and collide head-on.

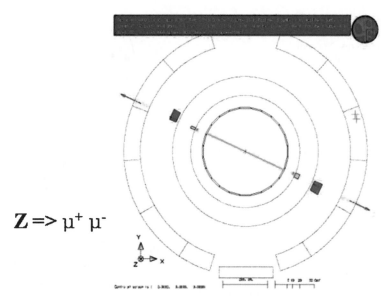

**Fig. 8.3** The decay of a Z particle into a muon–antimuon pair as observed by OPAL. The tracks of the two particles are registered in the central track chamber. The muons penetrate all the outer shells of the detector and energy bunches are deposited in the various calorimeters

The frequency of production of the Z particle and its decay into hadrons as a function of the collision energy is shown in Fig. 8.4. If the experimental points are fit by a theoretical formula the position of the resonance peak and the width of the resonance curve can be determined precisely. The results [1, 6–8] of the four experiments are in perfect agreement and by combining them one obtains for the mass of the Z particle the value $m_Z = (91.1875 \pm 0.0021)\,\mathrm{GeV}/c^2$ and for the width of the resonance $\Gamma_Z = (2.4952\pm = 0.0023)\,\mathrm{GeV}$. The resonance in the figure might seem quite wide; however, this is only so because the horizontal axis has been stretched enormously, with the zero point many metres to the left. The Z particle resonance is an extremely narrow line.

To achieve this incredible accuracy presented a considerable challenge both for the machine and for the experiments and only a close collaboration between physicists and engineers made it possible. The beam energies of about 50 GeV had to be determined and kept constant within a few megaelectronvolts, i.e. with a precision of about 1:5,000. As explained in Chap. 6, the beam energy is determined by the magnetic field along the beam path and by the diameter of the ring. Special measures can be taken to keep the magnetic field constant. However, the diameter of the ring is influenced by the sun and the moon, which produce tides not only in the ocean but also on the solid crust of the earth. In addition, rainfall changes the water levels in the Jura Mountains and at Lake Geneva, causing noticeable distortions of the ring. These effects led to changes of the ring diameter of the order of 1 m, producing measurable effects on the beam energy (see Fig. 8.5). When all these effects were understood and controlled, some erratic changes of the beam energy were still ob-

**Fig. 8.4** The Z particle production as function of the collision energy. Notice the extremely spread horizontal axis. The experimental points are the averages of the four experiments, with errors smaller than the points. The theoretical curves were calculated for different numbers of neutrino kinds

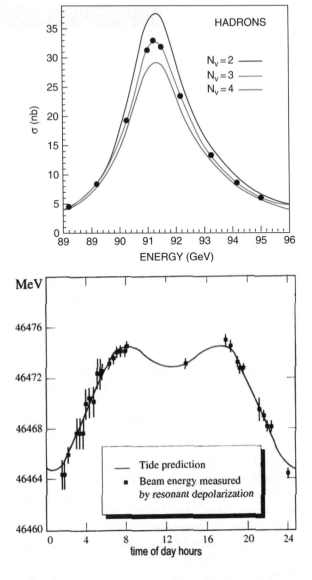

**Fig. 8.5** The influence of the tides on the LEP energy showing the influence of the moon and the sun. Notice the large expansion of the vertical axis

served and could not be explained for some time. Only when somebody compared the variations of the energy with the schedule of the fast French TGV train from Geneva to Paris could the mystery be solved. The motors of this train use direct current and for the return circuit the current is simply sent through the earth. However, instead of flowing through the ground the current preferred the LEP vacuum chamber, which had a much lower electrical resistance, producing an additional magnetic field, albeit rather small, but still changing the beam energy noticeably. To keep the beam energy constant was not sufficient; it had to be measured in absolute terms.

For this purpose several methods were developed, including the so-called resonant depolarization technique [9, 10]. Ultimately the residual systematic uncertainties in the beam energy could be limited to 0.0017 GeV for the mass of the Z particle.

## 8.6.1 Results for the Weak Interaction

From a precise measurement of the Z particle resonance one can deduce several fundamental results. The mass of the Z particle itself is a fundamental constant characterizing the weak interaction. It is remarkable that the Z particle is as heavy as a medium-heavy atomic nucleus, a surprisingly large mass for an elementary particle. Once the Z particle has been produced, it can decay in different ways. Mostly it disintegrates into hadrons and this decay mode has been used to determine the precise position of the resonance (Fig. 8.4). As already mentioned, the Z particle can also decay into a pair of charged leptons, $e^+e^-$, $\mu^+\mu^-$ and $\tau^+\tau^-$. The first two decay modes give very clean events in the detectors (Fig. 8.3), just two tracks of particles with opposite electric charges. The $\tau$ particle decays further and will provide interesting information on the strong force. All these decay modes can be observed separately. Within the experimental errors, no difference in the frequency of these three modes of decay into different lepton pairs has been observed. This is strong indication that the three kinds of charged lepton (e, $\mu$ and $\tau$) behave in the same way apart from their having different masses. This important observation is called 'lepton universality', implying that the electron, which played such a special role as a building block of matter for a long time, is just one partner among others.

The Z particle can also decay into a neutrino and its antineutrino, a decay mode which provides further fundamental information. Unfortunately such decays cannot be observed directly since the neutrinos leave the detector without a trace, but indirectly one can get the desired information. The total decay probability[17] of the Z particle is expressed by the sum of the individual decay channels, i.e. its total decay width $\Gamma_Z = \Gamma_{hadron} + 3\Gamma_{lepton} + \Gamma_{invisible}$, where $\Gamma_{hadron}$ is the decay width for the decay into hadrons which has been measured directly, $\Gamma_{lepton}$ is the decay width for the decay into charged leptons (assumed to be the same for e, $\mu$ and $\tau$, as confirmed by experiment) and $\Gamma_{invisible}$ indicates those decays which are invisible, i.e. for which the decay particles leave the detector without a trace. The latter one includes the decays into neutrinos or any other unknown particles with masses lower than half the mass of the Z particle.[18] The measurements gave

---

[17] According to the Heisenberg uncertainty principle, a fundamental law of quantum mechanics, the decay width is larger the shorter the lifetime of the particle, which in turn gives a larger decay probability.

[18] Because of the conservation of energy, the sum of the masses of the decay products cannot be higher than the mass of the Z particle.

$\Gamma_{hadron} = (1,744.4) \pm 2.0\,\text{MeV}$ and $\Gamma_{lepton} = (83.985 \pm 0.086)\,\text{MeV}$ and from these data one can calculate $\Gamma_{invisible} = (499.0 \pm 1.5)\,\text{MeV}$.

This last value hides essential information which can be extracted in the following way. The systematic character of the periodic system of the matter particles (Fig. 8.1) tells us that the weak charges of the first line of particles should be the same. With that knowledge one can calculate according to the standard model the decay width for the decay of the Z particle into a neutrino–antineutrino pair and it comes out as $\Gamma_{neutrino} = 167.23\,\text{MeV}$. Dividing $\Gamma_{invisible}$ by $\Gamma_{neutrino}$ provides us with a way to determine experimentally how many kinds of neutrinos $N$ exist in nature since $\Gamma_{invisible} = N \times \Gamma_{neutrino}$. The result is $N = 2.9840 \pm 0.0082$. This figure shows that indeed only three kinds of light neutrino as shown in Fig. 8.1 exist in nature. This is one of the most fundamental results obtained with LEP and is relevant not only for particle physics but also for astrophysics and cosmology. During the early stages of the universe many neutrinos were produced and these fill the cosmos as a kind of invisible gas playing a relevant role in the whole dynamics of cosmic development. Indeed they could be part of the mysterious dark matter, which is much more abundant than normal matter.

But a second basic conclusion can be drawn from this result. Since each neutrino is associated with a charged lepton, this implies that no more charged leptons exist and that the number of lepton families is also restricted to three. Because of the symmetry between leptons and quarks, it follows further that the scheme in Fig. 8.1 is complete and no more basic building blocks of matter fit into this scheme.

Of course, this does not forbid the existence of other particles not contained in the standard model framework, e.g. supersymmetric particles (see Chap. 14). These could have couplings different from those of the particles of the standard model and could lead to non-integer values of $N$. Hence, the measurement of $N$ with high precision has additional physical relevance since a deviation from an integer value would indicate the existence of such new particles. The quoted result is perfectly compatible with the standard model and does not give any hints for other particles. If they exist they must be heavier than the Z particle.

As mentioned above, the unification of the electromagnetic and the weak force in the standard model is not complete and one manifestation is the fact that the fundamental charges (coupling constants) of the two interactions appear in the standard model as independent arbitrary parameters which have to be determined experimentally. The elementary electric charge $e$ determines the atomic spectra, the observation of which provides a very precise value for $e$. Instead of $e$ one prefers to use the dimensionless 'fine structure constant' $\alpha = 2\pi e^2/hc = 1/137.035989$, where $h$ is Planck's constant and $c$ is the velocity of light. The fundamental weak charge which corresponds to $e$ is denominated as $g$ and the ratio between the two charges is called $\sin\theta = e/g$.[19] An important experimental task consists in measuring the

---

[19] For a detailed and precise analysis of the experimental data 'radiative corrections' have to be taken into account [1]. For this purpose an effective $\sin\theta$ is introduced, but these details cannot be considered here.

fundamental strength $g$, which in practice is equivalent to determining $\sin \theta$ since $e$ is very well known. This quantity cannot be deduced from the shape of the Z particle resonance but requires a different type of experiment i.e. asymmetries.

Various types of *asymmetries* can be observed:

- Forward–backward asymmetry: The number of fermions (e.g. muons) emitted in the forward direction relative to the direction of the incident electrons may be different from the number of the same kind of fermions emitted in the backward direction. The relative difference between these two numbers is called the "forward–backward asymmetry". It was measured at LEP 1 not only for all three kinds of lepton, but also for various heavy quarks.
- Left–right asymmetry: If the spins of the incident electrons (or positrons) are oriented parallel (antiparallel) to their direction of flight, they define a left (right)-handed screw. The production of a final state can depend on the handedness of the incident electrons. The change in counting rate associated with a flip of the spin direction in a particular final state gives the left–right asymmetry. Other more complicated asymmetries combining left–right with forward–backward asymmetries can be observed but are not discussed here. The Stanford Linear Collider (SLC)[20] at the SLAC in the USA could use polarized electrons and experiments at this collider achieved accurate measurements of such asymmetries which complemented the results at LEP 1 in an excellent way.
- Polarization asymmetry in the final state: Even if the incident particles are not polarized, the particles in the final state are polarized. This polarization cannot usually be observed except if the particle produced decays. Then its decay products exhibit a forward–backward asymmetry relative to their direction of flight. This is the case for the $\tau$ particle. It can decay into a number of various particles. The experiments at LEP 1 measured this asymmetry for five different decay modes of the $\tau$ particle and extracted a precise value for $\sin \theta$.

The combined results of these and further asymmetries obtained at LEP 1 gave a value of $\sin^2 \theta = 0.2324 \pm 0.0012$. This fundamental result fixes the ratio of the strength of the force between the electromagnetic and the weak interaction and tells us something about the strength of the weak force. Indeed, since $e^2 = g^2 \sin^2 \theta$, it follows that $g = 2.074e$. This is a very surprising result since it implies that the weak force is more than twice as strong as the electromagnetic force. Is, therefore, the name 'weak' interaction wrong?

This discrepancy can be understood if we take into account that the strengths of the different forces change with the interaction energy (see Sect. 8.7) and indeed

---

[20] The SLC was an electron–positron collider in competition with LEP 1. It could use polarized particles, but the number of Z particles produced was much lower than that at LEP 1. Because of the maximum energy of the Stanford Linear Accelerator Center (SLAC) linear accelerator and its smaller circumference, the SLC could not go to higher energies and could not compete with LEP 2. Combining the LEP 1 results with those obtained at the SLC, one obtains the more precise value $\sin^2 \theta = 0.23153 \pm 0.00016$.

the rate of change is different for different interactions. The weak nuclear force was discovered in β-decay of atomic nuclei involving energies (smaller than 1 MeV in most cases) about 100,000 times smaller than those available in Z particle decays (about 50 GeV). The effective strength of the weak force at low energies is determined by the quantity $G_F = \sqrt{2}g^2/8m_W^2$, where $m_W$ is the mass of the W particle and $G_F$ is the Fermi constant, as introduced and measured for nuclear β-decays. The apparent weakness of the 'weak' interaction and the smallness of $G_F$ stem from the large mass of the W particle. At high interaction energies as available at LEP the weak interaction is comparable to or even stronger than the electromagnetic force.

## 8.6.2 Results for the Strong Nuclear Force

Since electrons do not feel the strong force, being 'blind' to it because they do not carry colour charge, one might think that electron–positron colliders give mainly information on the weak force and that they can hardly contribute to elucidating the strong force. This understandable expectation is wrong as had been shown already by previous smaller electron–positron colliders. Hence, it is not too surprising that LEP could also make important contributions to the determination of fundamental parameters of the strong interaction.

One of the most important parameters is the strength of the force. To characterize its strength it has become habit to use, as for the two forces mentioned above, a dimensionless quantity $\alpha_s$. This coupling constant changes more rapidly with the interaction energy than for the other forces; thus $\alpha_s$ varies considerably even within the energy range covered by LEP. Hence, it becomes particularly interesting to measure this coupling strength as a function of the interaction energy.

The strong interaction could be studied at LEP by investigating the behaviour of the quarks which appear in the final state after the electron–positron annihilation. When two quarks are produced, it often happens that one of the quarks emits a gluon. This process is called 'gluon bremsstrahlung' since it is similar to the emission of a photon by an electron which is decelerated. As explained already, coloured particles do not exist as free particles in nature and hence each of the quarks and also the bremsstrahlung gluon fragment into normal (white) particles which form a jet of particles. Without gluon emission, the two quarks lead to events with two jets in the detector, whereas with the emission of a gluon a third jet appears. The gluon emission is determined by the strength of the strong interaction and therefore this can be deduced simply from the ratio of events with three jets to those with two jets. This possibility was used at PETRA at DESY in 1977 to prove the existence of the gluon [11]. However, at LEP the energies of the quarks and gluon produced were much higher and as a result the jets were much narrower and could be better isolated, providing better discrimination between the two kinds of event (see Fig. 8.6). The result of the LEP 1 experiments [12] for the coupling strength of the strong interaction is $\alpha_s(m_Z) = 0.119 \pm 0.003$, where the subscript $s$ indicates that the value refers to the strong force and $m_Z$ indicates that this value was obtained at the energy

**Fig. 8.6** A three-jet event
observed by L3. Two jets
arise from the fragmentation
of quarks and one from the
fragmentation of the gluon

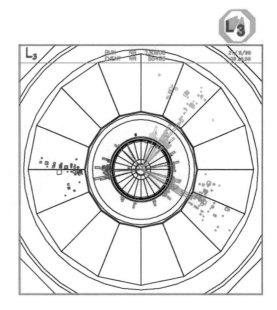

of the Z particle resonance. $\alpha_s$ can also be determined at other energies around the Z particle peak, and the results obtained are shown in Fig. 8.7.

However, it is quite surprising that most of the values and the best values for $\alpha_s$ at rather low energies were also obtained at LEP. To get results at low energies the three-jet events cannot be used, and another method has to be applied. This opportunity arises thanks to the fact that quite a number of $\tau$ particle pairs are produced in Z particle decays which are well isolated from other events and whose decays can be investigated in detail. The rate of $\tau$ particle decays into hadrons is proportional to $\alpha_s$ and the observation of such decays has given the value [13, 14] $\alpha_s$ (1.777 GeV) $= 0.345 \pm 0.010$, where 1.777 GeV corresponds to the mass of the $\tau$ particle. Thus, the LEP results alone, combining data from low up to the highest energies obtained thus far (see Fig. 8.7), show that the strong coupling $\alpha_s$ changes fast with energy as predicted by quantum chromodynamics (QCD). This is one of the most important tests of the theory for the strong force.

The gluon carries a colour charge, which is a special property of the strong force and permits a direct interaction between gluons, a circumstance originating in the so-called non-Abelian character of the SU(3)symmetry group. Without going into details, it may be mentioned that the observation at LEP of four-jet events and their properties provided evidence for the direct interaction of gluons [15, 16], another important result to confirm QCD.

Many other measurements confirming QCD or giving additional information cannot be mentioned here.

**Fig. 8.7** The coupling strength of the strong nuclear interaction as a function of the interaction energy ($E_{cm}$). The *circles* represent data from the JADE experiment at the PETRA collider at DESY and the *squares* represent data from LEP experiments. The *line* is a theoretical fit (quantum chromodynamics, *QCD*) to the data. It is clearly shown that $\alpha_s$ is not a constant but changes with energy ('running')

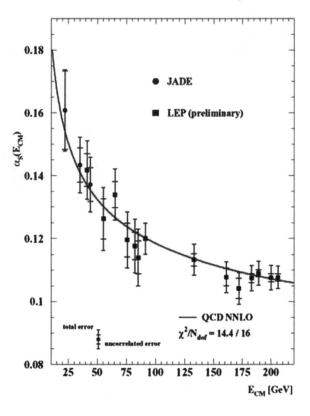

## 8.7  Results from LEP 2

### *8.7.1  W Particle Production*

After the Z particle had been extensively studied, the next task for LEP was to investigate the second carrier of the weak force, the W particle. Because W particles are electrically charged, they cannot be produced singly as can the neutral Z particles. Since electron–positron annihilation leads to an electrically neutral state and because of the conservation of the electric charge, the final state must contain a pair of particles with opposite electric charges, e.g. a $W^+ W^-$ pair. The collision energy must be sufficiently high to produce at least two of these heavy particles (any additional particles with a total electric charge of zero require extra energy for their production). To achieve this goal, the beam energies of LEP were successively increased during its second phase of operation (see Table 14.1) and in 1996 the threshold energy for the production of W particle pairs of about 80 GeV per beam was reached. Above this threshold the production of single Z particles competes with the production of W particle pairs. It is relatively easy to separate the two kinds of process.

As has been shown the Z particle decays into a pair of particles of the same kind, e.g. $e^+e^-$ or $\mu^+\mu^-$, which fly apart in opposite directions (see Fig. 8.3). The W particle decays preferentially into a charged lepton and a neutrino,[21] which can only be detected indirectly by measuring the missing energy and momentum which the neutrino carries away. The simplest types of W particle pair events are those where each W particle decays into a lepton and a neutrino, which gives in the detector combinations of $e^+e^-$ or $\mu^+\mu^-$ or $e^+\mu^-$, etc. However, here the two leptons do not fly apart in opposite directions as in the case of Z particle decays, but form various angles indicating missing energy and momentum. It is easiest to single out events with $e^+\mu^-$ since they cannot originate from Z particle decays (Fig. 8.8). In Fig. 8.9 the production of single Z particles and of W particle pairs over the whole energy range investigated is shown. Most of the data came from LEP experiments. At about 160-GeV collision energy, where sufficient energy is available to produce W particle pairs, this process becomes more frequent than single Z particle production.

Since the threshold energy corresponds to just twice the mass of the W particle, observation of the W particle provides a value for its mass. To determine the precise mass of the W particle is, however, more difficult than the measurement of the mass of the Z particle, in which case just changing the collision energy was successful. Here two W particles are produced and therefore the events are more difficult to analyse. The $W^+$ particle can also decay into an up quark and an anti-down quark (respecting the conservation of charge) or a charm quark and an anti-strange quark[22] and these quarks can be detected as hadronic jets.

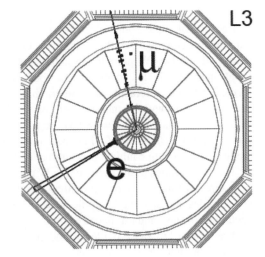

**Fig. 8.8** An event with the production of a W particle pair as observed by L3. The W particles decay alternatively into an electron and a muon. The associated neutrinos can be detected only indirectly by missing energy and momentum

---

[21] Electric charge must again be conserved; hence, the final state must have the same charge as the decaying W particle, e.g. $W^+ \rightarrow e^+ + \nu$.

[22] It cannot decay into a top and an anti-bottom quark, since as we know now, the top quark is much heavier than the W particle.

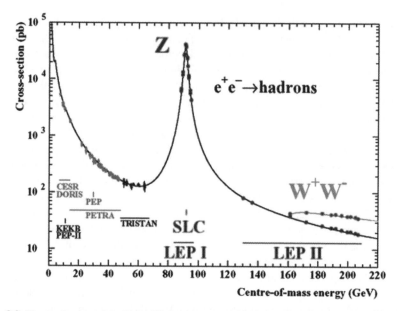

**Fig. 8.9** The production of the Z particle in electron–positron annihilation over the entire energy range investigated. The decay of the Z particle into hadrons is registered. The resonance at 91 GeV dominates. At about 160 GeV, the threshold for the production of W particle pairs is crossed and this process becomes more important than single Z particle production. The data were obtained by experiments at various colliders, but most of the data originated at LEP

Because of these complicated decays it is not possible to observe a clear W particle resonance as for the Z particle. One has to study the properties of the decays of the two W particles in detail and infer their mass by a more complicated analysis which cannot be described here. All four LEP experiments performed such measurements and the combined results [17] including leptonic and hadronic decays and the measurement of the threshold energy give[23] very precise values for the mass of $m_W = (80.376 \pm 0.033)\,\text{GeV}/c^2$ and for the decay width of $\Gamma_W = (2.196 \pm 0.083)\,\text{GeV}$. Thus, two more fundamental constants characterizing the weak interaction could be measured at LEP.

The probabilities of the decay of the W particle into the individual leptonic channels agree within the errors, providing another test of lepton universality. These probabilities and also the hadronic decay width are in full agreement with the expectations of the standard model.

The standard model provides a fundamental relation between the masses of the Z and W particles, namely $m_W/m_Z = \cos\theta$. Using the results from LEP 1 for the Z particle resonance (above all the measured values of $m_Z$ and $\sin\theta$, but also other results), one can use this relation to predict the mass of the W particle and one

---

[23] The W particle could later also be produced at the Tevatron proton collider in the USA and a mass of $(80.429 \pm 0.039)\,\text{GeV}/c^2$ was found, confirming very nicely the LEP result.

**Fig. 8.10** The production of W particle pairs as a function of the collision energy. The standard model calculation based on the assumption that the W and Z particles carry weak charges and hence 'feel' a direct force is in excellent agreement with the experiments. Two theoretical curves calculated with the assumption that the W and Z particles carry no weak charge lie far above the experimental data

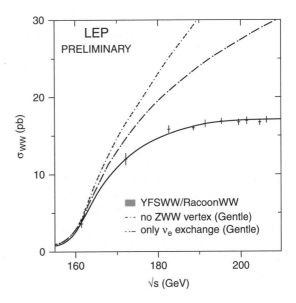

obtains[24] $(80.363 \pm 0.032)\,\text{GeV}/c^2$. This is in excellent agreement with the directly measured value and it provides one of the most significant tests of the standard model.

The LEP experiments could carry out another fundamental test of the standard model, which predicts that the Z and W particles carry weak charges and therefore can interact directly. This has important consequences for the production frequency of W particle pairs as a function of the collision energy. A comparison between the LEP 2 results and the theoretical expectations [2–4] is shown in Fig. 8.10. It can be clearly seen that theories without a direct ZWW coupling cannot reproduce the data.

## 8.7.2 Looking for the Invisible – the Top Quark

When LEP was conceived there was the hope that its collision energy would be sufficient for the production of the yet unknown top quark and some experiments were designed to study in detail the decays of this unconfirmed quark. However, no theory could make any predictions for the mass of the top quark and the search for it was a shot in the dark. Unfortunately it turned out that the top quark was too heavy to be observable at LEP. However, sometimes it is possible in physics to get indirect information on 'invisible' objects by observing tiny perturbations of otherwise well understood phenomena. A well-known example is the prediction of the existence of

---

[24] To calculate this parameter radiative corrections have to be taken into account as explained in the next chapter.

new planets (e.g. Uranus) from small perturbations of the orbits of known planets. When more powerful telescopes became available, the predicted planets could be observed directly and the predictions were verified.

This is exactly what happened at LEP with the top quark. From the extremely precise measurements at LEP the mass of the top quark could be determined even though it was not directly observable. Later the top quark was discovered at the more powerful Tevatron proton collider in the USA and the mass determined by LEP was confirmed with great accuracy. The details of this exciting story cannot be reported here, but at least the principle may be explained.

According to the Heisenberg uncertainty principle of quantum mechanics, it is possible to produce particles 'from nothing', i.e. without the necessary energy supply, albeit only for a short time. One can borrow energy, so to speak, from a vacuum, but this has to be returned very fast, however. The more energy one needs (i.e. the heavier the particle to be produced), the shorter is the time one can have it.[25] This effect plays a role in all kinds of reactions between particles at high energy. In the electron–positron annihilation process, the incident particles or the particles produced after the annihilation can emit a particle even if sufficient energy is not available, provided it is reabsorbed after such a short time that the Heisenberg uncertainty principle is respected. Such processes are typical for quantum field theories and are called 'radiative corrections' and the temporarily emitted particles are called 'virtual'. However, the effects are small and can only be detected with precision measurements.

In quantum electrodynamics (QED), the field theory for the electromagnetic interaction, an electron can emit a photon, which is immediately absorbed again. Such processes influence the energy levels of atoms and can be detected by observing spectral lines. When such a level shift due to radiative corrections was first observed (the 'Lamb shift'), it confirmed the theory of QED and resulted in the award of the Nobel Prize to both the theorists and the experimenters. The precision experiments at LEP made it possible to measure similar radiative corrections for the electroweak theory, thus confirming with a precision of better than 1% the standard model. Again a Nobel Prize was awarded, but this time only to the theorists since too many experimenters were involved.

Apart from confirming the standard model, the observation of radiative corrections provided additional interesting information. One of the processes contributing to radiative corrections is the emission of a top quark. Although it was too heavy to be produced at LEP, it plays a relatively large role as a virtual particle since theory tells us that the radiative corrections due to the top quark are proportional to the square of its mass, $m_{top}^2$. Since the mass of the top quark $m_{top}$ is rather large, these effects, although still tiny, could be measured and the unknown mass of the top quark was deduced without having produced real top quarks. It was possible to determine the mass of the top quark rather accurately by fitting all the precision

---

[25] Quantitatively the uncertainty principle is $E.t = h/2\pi$, where $E$ is the energy available for the time $t$ and $h$ is Planck's constant.

results at LEP 1 and a value of $m = 173^{+13}_{-10}$ GeV/$c^2$ was found. It was a real triumph when much later the top quark was produced at the Tevatron and its direct mass measurement agreed beautifully with the prediction from the LEP experiments. The most recent combined results [2–4] from the two experiments at the Tevatron give $m_{\text{top}} = 170.9 \pm 1.8$ GeV/$c^2$.

### 8.7.3 The Higgs Particle – Disappointment but!

One of the main shortcomings of the standard model is the unexplained masses of the Z and W particles. If the interaction strictly obeyed the symmetry SU(2)×U(1), then these carriers of the force would have no mass at all, like the photon. To repair this deficiency a mechanism was invented, the so-called spontaneous symmetry breaking explained in Sect. 8.4. Such a spontaneous symmetry breaking is introduced into the standard model by assuming the existence of an additional field filling the whole space whose field quanta have spin 0, the only particle in the standard model without spin. Since such a mechanism was proposed for the first time by Peter Higgs, and the field and the field quantum carry his name. Whether such a mechanism can reproduce the experimental data or not is one of the central questions in particle physics today. The existence of a Higgs field would not only explain the masses of the Z and W particles, but also those of all other particles, such as leptons and quarks. The experimenters have a hard time since theory can give no indication at all in what range the mass of the Higgs particle may be found. At each new accelerator that came into operation the Higgs particle was looked for but without success. Of course, everybody was hoping that at last at LEP the Higgs particle could be produced and its properties studied and this search became an exciting story.

Since no Higgs particle was found in the decays of the Z particle, the Higgs particle must be heavier. The Higgs particle will decay preferentially into the heaviest particles allowed by energy conservation and hence its decay products will be preferentially bottom quark pairs, or Z particle pairs, if heavy enough, or even top quark pairs. But in that case the Higgs particle would be too heavy to be produced at LEP. At LEP 2 a hectic search started and lasted until its last days whenever a higher beam energy was reached. These events will be described in more detail in Chap. 14. The search for the Higgs particle was unsuccessful, but at least a lower limit for its mass, independent of theoretical assumptions, could be established: $M_{\text{Higgs}} \geq 114$ GeV/$c^2$. This is so far the most reliable information one has about the Higgs particle.

But this was not the end of the search for the Higgs particle at LEP. Like the top quark, the Higgs particle can also participate in the annihilation process as a virtual particle. Unfortunately theory tells us that the radiative corrections due to the Higgs particle are relatively small: they are proportional only to $\log M_{\text{Higgs}}$. Nevertheless, by combining all the precision measurements (including the most recent values for the mass of the top quark) and making a fit with the mass of the Higgs particle

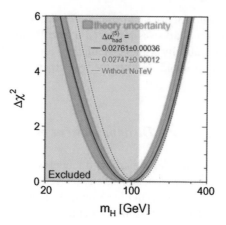

**Fig. 8.11** The probability of finding a certain value for the mass of the Higgs particle derived from radiative corrections. The *curves* were calculated for different values of the strong coupling constant $\alpha_s$. There is a remaining theoretical error indicated by the *band* near the theoretical curves. Surprisingly, the expected mass of the Higgs particle is low, around $100 \, \text{GeV}/c^2$. The mass range excluded by the direct search for the Higgs particle is shown by the *shaded area*

as a free parameter, one can get a probability estimate for its value. Surprisingly it turns out (see Fig. 8.11) that the most likely mass of the Higgs particle is around $100 \, \text{GeV}/c^2$, with an upper statistical limit of about $200 \, \text{GeV}/c^2$. Of course, the mass cannot be lower than the direct lower limit of $114 \, \text{GeV}/c^2$, but this analysis indicates that the Higgs particle should be rather light and it should certainly be in the reach of the next generation of colliders, i.e. the LHC or even the Tevatron might have a chance of finding it (see Chap. 14).

## 8.7.4 Hints Beyond the Standard Model

The ultimate goal of fundamental physics is to reduce all natural phenomena to a set of basic laws and a theory that, at least in principle, can quantitatively reproduce and predict the experimental observations. The standard model does not achieve this in spite of its great success. As already mentioned, the standard model does not fully unify the three interactions, electromagnetic, weak and strong, it does not explain the strength of the forces and the masses of the particles, i.e. leptons, quarks, W and Z particles, and these have to be introduced as free parameters, according to the counting more than 20.

Therefore great efforts have been made by theorists to embed the standard model into a wider theory avoiding the deficiencies of the standard model. The most popular extension of the standard model is the supersymmetry theories. The various versions cannot be explained here but the principle is the following. We have seen that a symmetry exists between quarks and leptons, both families being grouped in three doublets which correspond to each other. Both quarks and leptons have spin $^1/_2$; they are fermions. On the other hand, the carriers of the forces are particles with spin 1; they are bosons. In supersymmetry theory a further symmetry is postulated: a symmetry between constituents of matter (quarks and leptons) and the particles transmitting the forces (field quanta as carriers). It is assumed that each 'normal' particle has a 'supersymmetric' partner which has a spin differing by 1/2. For the matter particles the partners are indicated by a leading 's', whereas the partners of the force carriers are indicated by '-ino'. Hence the quarks would have partners called 'squarks', with spin 0, and similarly leptons would be accompanied by sleptons, with spin 0. The partners of the photon, gluon, W and Z particles are denominated 'photino', 'gluino', 'Wino' and 'Zino', all having spin 1/2. Even the Higgs particle gets several companions called 'shiggs', with spin 1/2, and some of them can even carry an electric charge.[26] In this way a supersymmetric world is postulated which complements our ordinary world. None of these particles had been detected and the question whether this supersymmetric world exists or not was one of the exiting challenges for LEP.

The theory predicts that some of these supersymmetric particles should be relatively light and hence there was the tantalizing hope that they could be detected at LEP. Nature, however, had decided otherwise. The searches for these and other exotic particles were unsuccessful, except that (as for the Higgs particle) lower limits for the masses of these particles could be established in the region from 50 to 70 GeV/$c^2$.

So far there exists only one indication for phenomena beyond the standard model and it comes from the precise LEP experiments. An ultimate goal of particle physics would be to unify the three forces of the microcosm into one single fundamental force. Several theories have been developed for this aim, called grand unification theories (GUT) and some versions of supersymmetry theory belong to this class of theories. They predict that the coupling strengths $\alpha_i$ of the forces change with the energy $E$ at which the interaction takes place and one talk about 'running coupling constants' (see Sect. 8.6.2). The theory predicts that they change according to the formula $\alpha_i = 1/N_i \log E$, where the subscript $i$ indicates the particular force and the constant $N_i$ depends on the number of particles which 'feel' the force at the energy $E$. The formula suggests a plot of $1/\alpha_i$ as a function of $\log E$ should give straight lines whose slopes are determined by $N_i$. If the three forces can be united in a fundamental single force one would expect that at a certain energy all forces should have the same strength, implying that all three $\alpha_i$ plotted against the energy $E$ should meet at one point.

Before LEP it seemed indeed that this condition was met within the relatively large errors prevailing at that time. When the LEP experiments provided much more accurate data for the $\alpha_i$ it turned out that the extrapolations for the running $\alpha_i$ did not come together at the same point (Fig. 8.12, top). However, by assuming that supersymmetric particles exist, one can get the three $\alpha_i$ to meet at one point (Fig. 8.12, bottom) under the condition that the supersymmetric particles have

---

[26] The gravitational force, which is too weak to play a role in particle physics and hence is not considered here, is transmitted by still hypothetical gravitons, field quanta with spin 2. They would have a supersymmetric partner, the gravitino, with spin 3/2.

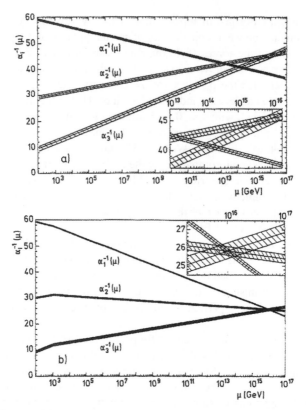

**Fig. 8.12** The inverse of the coupling strengths $1/\alpha_i$ as function of $\log \mu$, where $\mu$ is the interaction energy. According to theory, the functions should be straight lines. In the case of unification of the forces they should intersect at one point. The precision of the LEP experiments indicates that in the framework of the standard model this is not the case (*top*). Assuming supersymmetry and masses of the supersymmetric partners around $2\,\mathrm{TeV}/c^2$, one can obtain an intersection at one point (*bottom*). The *insert* shows a blow-up of the intersection points

masses around $2\,\mathrm{GeV}/c^2$. This is in agreement with the fact that they have not been found so far at lower energies. This behaviour of the coupling constants is the only indication so far of a deviation from the standard model. All other experimental data at LEP and also in other experiments are reproduced perfectly by the standard model.

## 8.8 Summary of LEP Results

The most important results of the LEP experiments may be summarized in the following way:

1. The experimental determination of the fundamental parameters of the electroweak force, including the properties of the W and Z particles (the carrier of the force) and the strength of the force (its coupling constant).

2. The proof that only three kinds of (light) neutrino exist in nature. This shows that the periodic scheme of the buildings blocks of matter is complete. This result is also relevant for astrophysics.
3. A number of precision tests of the validity of the standard model of particle physics.
4. The proof that the standard model is a renormalizable quantum field theory, a result worth the award of the Nobel Prize to the theorists.
5. Showing that the strength of the strong nuclear force is changing with the interaction energy ('running coupling constant'), verifying the prediction of QCD, the theory of the strong interaction.
6. The indirect determination of the mass of the top quark, facilitating its discovery by the Tevatron collider, which confirmed excellently the prediction.
7. A lower limit for the mass of the hypothetical Higgs particle could be given, the most solid existing information for this particle. Limits for supersymmetric and other hypothetical particles were also obtained.

As mentioned earlier, one of the reasons for approving four LEP experiments was the hope that by combining their results, we could improve the overall accuracy. Indeed at LEP it happened for the first time that the experimental teams collaborated in working groups comparing and combining their data. This was a unique feature and a milestone in scientific cooperation.

This summary cannot, of course, give appropriate credit to many additional detailed results. Although no real surprise was produced by the LEP experiments and few headlines appeared in public journals, our concept of the microcosm has nevertheless changed through our obtaining a surer basis. Certainly LEP has changed high-energy physics from a '10% accuracy science' into a precision field. A number of theories could be discarded and, on the other hand, precious indications for the direction of further research were obtained. In particular, the construction of the LHC and its parameters could be justified.

# References

1. Grünewald MW (2008) Experimental precision tests for the electroweak standard model. In: Elementary particles, vol 1. Landoldt-Börnstein, Heidelberg
2. 't Hooft G (2007) The making of the standard model. Nature 448:271
3. Wyatt T (2007) High-energy colliders and the rise of the standard model. Nature 448:274
4. Altarelli G (2008) The standard model of electroweak interactions. In: Elementary particle physics, vol 1. Landoldt-Börnstein, Heidelberg
5. Brown D (2000) Angels and demons. Pocket Books, New York
6. ALEPH, DELPHI, L3, OPAL, SLD Collaborations (2006) Precision electroweak measurements on the Z resonance. Phys Rep 427:257
7. ALEPH Collaboration, DELPHI Collaboration, L3 Collaboration, OPAL Collaboration, LEP Electroweak Working Group, SLD Electroweak, Heavy Flavour Groups (2003) A combination of preliminary electroweak measurements and constraints on the standard model. Available via arXiv. http://arxiv.org/abs/hep-ex/0312023

8. LEP Electroweak Working Group (2003) The LEP Electroweak Working Group. http://lepewwg.web.cern.ch/LEPEWWG
9. Brandt D, Burkhardt H, Lamont M, Myers S, Wenninger J (2000) Rep Prog Phys 63:R1–R62
10. Assmann RW, Lamont M, Myers S (2002) A brief history of the LEP Collider. CERN report CERN-SL-2002-009-OP. CERN, Geneva
11. Burrows PN (1988) Particles Fields 41:375
12. Kluth S (2006) Rep Prog Phys 69:1771
13. Davier M, Hocker A, Zhang Z (2006) Rev Mod Phys 78:1043
14. Cavier M, Hocker A, Zhang Z (2006) The physics of hadronic tau decays. Available via arXiv. http://arxiv.org/abs/hep-ph/0507078
15. Ali A (1986) In: Physics at LEP. CERN 82-2, vol II. CERN, Geneva, p 109
16. Rudolph G (1986) In: Physics at LEP. CERN 82-2, vol II. CERN, Geneva, p 150
17. ALEPH Collaboration, DELPHI Collaboration, L3 Collaboration, OPAL Collaboration, LEP Electroweak Working Group (2006) A combination of preliminary electroweak measurements and constraints on the standard model. Available via arXiv. http://arxiv.org/abs/hep-ex/0612034

# Chapter 9
# Creating New Technologies

The main objective of a laboratory such as CERN is the accumulation of new fundamental knowledge and the development of scientists and engineers, whereas the development of new applications does not belong to its formal mandate. Usually the *results* of basic research themselves do not lead immediately to new applications and hence to improvements in social welfare. However, history has shown that the discovery of fundamentally new phenomena results in the long run in completely new technologies. Furthermore, there is another way in which activities like those at CERN can lead to immediate technical spin-offs. To penetrate into the microcosm, new instruments have to be developed which require new technologies. Hence, CERN had to acquire an enormous technical competence in many fields and the resulting procedures which led to positive economic consequences will be considered. The most impressive example is the invention of the World Wide Web at CERN.

## 9.1 Basic Research Leads to Quantum Jumps in New Technologies

Although fundamental new discoveries often seem to have no practical applications, in the long run they may give rise to completely new technologies. The discovery of electricity gave us electric light. If a government had started an innovation programme in the nineteenth century, it would probably have concentrated on the development of better gas lamps. The unification of electricity and magnetism in the nineteenth century formulated in Maxwell's equations was a contribution to pure fundamental knowledge, but in the twentieth century it became the basis of electric motors and generators and thus of the whole electrotechnical economy. Many other examples could be given.

In a more systematic way, I tried to demonstrate [1–3] how basic physics research eventually results in unexpected applications (see Fig. 9.1). From human dimensions (about 1 m) physics penetrates, on one hand, into the microcosm by investigating ever smaller objects, from normally sized objects to atoms, atomic nuclei, hadrons such as photons and neutrons and finally quarks (Fig. 9.1, left). On the right side of Fig. 9.1, the exploration extends from Earth to planets, stars, galaxies, groups

H. Schopper, *LEP – The Lord of the Collider Rings at CERN 1980–2000*,
DOI 10.1007/978-3-540-89301-1_9, © Springer-Verlag Berlin Heidelberg 2009

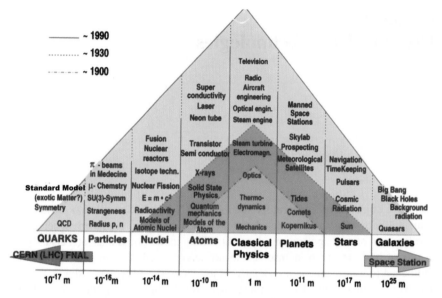

**Fig. 9.1** Pyramid showing the growth of applications based on fundamental science. Along the base the size of the objects investigated is shown: to the *left* penetration into the microcosm, to the *right* exploration of the universe. The different pyramids indicate approximately the state of our knowledge at various times

of galaxies, quasars and black holes. If the plot is made on the basis of a pyramid, one can show how applications developed by building up the pyramid from the bottom to the top. Basic physics has always been in the outmost corners at the base of the pyramid, at each time apparently far removed from any useful application. However, when the pyramid grew, the basic results became the foundation of revolutionary applications. At the same time, the 'abstract' fundamental knowledge became familiar. When switching on an electric light or watching TV, people have the impression that they are familiar with the electromagnetic phenomena although if they are asked what carries the electromagnetic waves through empty space, only a few of them would be able to give the correct answer. There is no logical reason why the pyramid should not continue to grow, although it is impossible to predict the speed.

I would like to give just one example of how extremely abstract ideas led to new applications. John Bell, a theoretician at CERN, developed one of the most fundamental theories [4] and proposed experiments which would allow decisions to be made on the various interpretations of quantum mechanics. Einstein thought that the probability interpretation of quantum phenomena is wrong ("God does not play dice") and that some 'hidden parameters' must reestablish classical causality. The experiments [5–7] proposed by Bell involving so-called entangled states were carried out and it was shown that Einstein was wrong. Unfortunately Bell, a charming, soft-speaking Irishman, died too early in 1990 at the age of 62, otherwise he would certainly have been awarded the Nobel Prize. Experiments with entangled states have become the basis for a new technology of transmitting secure coded messages because any interference with the message by an eavesdropper will reveal his presence. Several commercial firms now offer such systems and one was used for the first time in 2007 to transmit the results of a public vote in Geneva. Entangled states may also become the essential elements of quantum computers.

Can it definitely be excluded that the unification of the weak and electromagnetic forces, one of the main achievements of modern particle physics, might one day find unexpected applications? The results of basic research provide not only the basis for new technologies, but they are also in more general terms the prerequisite to keep open options and possible scenarios for decisions of governmental and other authorities.

## 9.2 The Technological 'Spin-Off'

As mentioned in Sect. 9.1, new technologies depend on fundamental research. But a similar relation also exists in the opposite direction. Research is often limited by what is technically available. To be able to make progress, fundamental sciences need new instruments. The use of the telescope enabled Galileo to revolutionize the view of the universe and the microscope, invented at about the same time, offered new aspects of the microcosm and became essential for biology and medicine. To develop new experimental techniques, close cooperation between scientists and engineers is often required. In many cases scientists have been instrumental in pushing back technical frontiers, and engineers have helped to provide scientists with new tools.

Such cooperation very often gives rise to what is called 'spin-off' or 'fall-out'. This is particularly true for a laboratory such as CERN which has the task to provide to the outside users from national laboratories and universities the most modern facilities for research. Forefront scientific research can only progress if it is coupled with technological innovations. To cope with this challenge the CERN staff have acquired over the years a remarkable technological competence, which indeed has become a basic ingredient of the success of CERN. As mentioned in Chap. 6, the construction of LEP presented many technical challenges and coping with them led to many spin-offs. This is also true for the development of the LEP detectors, which was done both inside and outside CERN.

The mechanisms by which technology transfer can be achieved can be quite different and these are discussed in the following sections.

### 9.2.1 Transfer Through Patents and the World Wide Web

CERN can take out a patent on a new development which then could be exploited by industry. This method of technology transfer works very well for institutions which have the genuine task of technical innovation. Fundamental research as carried out at CERN depends very strongly on an atmosphere of open exchange and full and rapid publication of results. Therefore, CERN had not placed strong emphasis on taking out patents, which unfortunately sometimes hinders spin-off. I remember a visit to CERN by high-level industrial leaders who at the end of their visit told me that they had seen many interesting developments and they asked how they could benefit from the corresponding patents or trademarks. When I answered that there were no such patents they completely lost interest since industry is preferably interested in protected know-how. To remedy this situation CERN has in the meantime established a special department for technology transfer and a patent policy.[1]

There was one case where the renouncing of taking out a patent was greatly regretted. The World Wide Web was invented at CERN in 1989. The need to make the LEP results available to the many hundreds of outside users was met by developing the World Wide Web. In the meantime, the Web has conquered the world, has changed the daily lives of many people in industrialized countries, and is penetrating into more domains of public life – maybe to the regret of some people. One can hardly think of any other spin-off from basic research which has had similar consequences. Each time one types 'http' or 'www' one unknowingly pays tribute to CERN. Quite often we were asked why CERN did not make money from it. Unfortunately, like for other inventions[2], we did not recognize its relevance for the Internet and did not think of protecting this innovation. Following usual practice, the organization left it to the inventors to exploit their invention. Thus, Tim Berners-Lee, the main inventor,[3] contacted the EU but did not find any interest. Eventually he left CERN and formed the World Wide Web Consortium in September 1994. Fulfilling the needs of the CERN users has led to one of the most important CERN spin-offs.

## 9.2.2 Joint-Development Contracts

As explained in Chap. 6, certain components were needed for LEP which could not be found on the market. In a number of cases industry did not have the necessary know-how to produce them or did not want to acquire it at its own cost. It is common practice in other fields to sign a development contract with industry which is remunerated according to the expenses incurred. Since industry usually does not want to take any risk, such contracts are relatively expensive and it is difficult to control the delivery time and the final specifications. In addition, one has to make sure that for the delivery of consumables a firm does not become the only supplier, because it can then dictate the prices.

---

[1] See http://ttpromo.web.cern.ch/ttpromo/Home.do; http://ref.web.cern.ch/ref/CERN/HR/AT2002/O1/

[2] At the beginning the Xerox copying system was also considered to have no chance on the market.

[3] Robert Cailliau was the other inventor, but he did not like the limelight so much. He is still at CERN.

In such cases CERN very often signs an agreement with a company to develop in a common effort a product which does not yet exist. In this way the relations between CERN and industrial firms become quite different from a straight purchase or even development contract. A vendor–customer type of contract becomes a collaboration. The realization of such a collaboration can be quite different depending on the specific subject. In a few cases it even happened that CERN acted as a catalyst by bringing competing firms together, either to collaborate or at least to agree on common standards. A few examples may give a clearer picture.

In the vacuum chamber of the LEP main ring an extreme vacuum had to be created and maintained during operation, even with considerable outgasing due to the synchrotron radiation hitting the walls of the vacuum tube. A powerful pumping system was needed which at the same time had to be cheap since it had to cover the whole ring of 27-km length. As mentioned in Chap. 6 a special non-evaporable getter material was developed in cooperation with industry. It is deposited on metal strips and absorbs the gases. The non-evaporable getter material resulted in two patents from CERN. One was exploited by the SAES Getter Group, which announced the branding of a new product under a licensing agreement,[4] and another patent developed by CERN staff member Cristophoro Benvenuti became the basis for a large solar plant realized by the European Council for Nuclear Research[5] with the firm SRB Energy.

Another example is provided by the klystron, an electronic device which provided the rf power for the accelerating fields in the LEP cavities. Such klystrons were on the market but only for pulsed operation, whereas for LEP such tubes were needed for continuous operation and with a power output of 1 MW and high efficiency. Such klystrons have a limited lifetime and their replacement or repair forms a non-negligible part of the running cost of the machine. To have more than one supplier and to reduce the risk that our specifications could not be met, we were prepared to adjudicate several development contracts. After a worldwide tendering procedure during which one firm declared it was not able to meet the specifications, two contracts were signed with two European firms. They both performed well and could later use the acquired know-how to sell such klystrons for other purposes.

Quite often CERN is used as a test-bed for new complex products. Failures in the initial stage do not lead to economic disasters and CERN, being a very intelligent customer, can even help to recognize and eliminate the bugs. An excellent example was the development of a local area network in cooperation with IBM in order to provide appropriate communication channels for a number of purposes inside CERN. Because it was a new product, before introducing it to the market, IBM wanted to test it with intelligent users who in the case of initial troubles would help to expunge them (Sect. 6.5.2).

---

[4] See http://saesgetters.com

[5] http://www.ingegneriasolare.org/index.php?option=com_content&task=view&id=173& Itemid=38; http://www.askeu.com/default.asp?ShowNews=3183

The development of detectors and the associated electronics gave rise not only to new applications, but has also led to the founding of some start-up companies. For example, the development of a special silicon readout chip for DELPHI resulted in a Norwegian firm which now sells self-triggering electronics for imaging applications in physics, space and biomedicine [8]. The experience with special detectors and the associated data handling has also given rise to cooperation with the Hôpital Cantonal at Geneva for the improvement of positron emission tomography diagnostics.

It might be mentioned here that some firms which received a development contract were asked to install themselves temporarily in a technical zone on the French side close to CERN. This happened for products which were difficult to transport such as the concrete magnets for LEP. The hope was that the temporary installation would lead to a technology park and indeed that is what happened.

## 9.2.3 Technology Transfer by Procurement

Another and very efficient way to transfer know-how is by contracts for the fabrication of technologically advanced components.

In cases where CERN had technical know-how in a specific field superior to that of industry, the following procedure was sometimes adopted. CERN developed a new technology and built prototypes to demonstrate the feasibility. To produce the components in larger quantities, industrial firms were then asked to bid for the contract, and the firm which submitted the best offer (best price, necessary competence and stability guaranteed) would get the contract. With the contract the CERN know-how would be made available to the firm. During the execution of the contract, continuous contact between CERN experts and staff from the firm would ensure that the specifications prepared by CERN engineers were met and in this way all the know-how, even if not documented, acquired at CERN was transferred to the firm. Using this know-how, the firm might then even apply for patents. Usually it was agreed in such cases that CERN could continue to benefit from the know-how without paying royalties.

One should, however, be aware that the main part of the economic benefit provided by CERN contracts does not appear immediately as a patentable idea or by the direct commercialization of a product. The process is much more indirect. It occurs through acquired know-how in advanced technology fields leading to decisive quality or performance improvements or by the opportunity given to enter new markets. There is also the marketing value of having been able to satisfy a demanding customer such as CERN.

The benefits a firm can obtain in this way are wide-ranging. In some cases the firms acquired a new technology which increased their general turnover. Examples can be found in many fields, in particular in electronics. In other cases it was a matter of learning how to respect tolerances. A small British firm which in the past had produced simple metal-sheet furniture learned to fabricate the rf wave guides for LEP. These were not so different from simple metal boxes except that their dimensions have to follow critical tolerances. After the execution of the CERN contract, the firm was suddenly in the high-tech market. For the installation of the LEP machine and for the detectors, large, heavy pieces weighing many tons had to be moved with a precision of millimetres. A shipbuilding firm developed the necessary technical equipment, which later turned out to be useful in shipyards. CERN is known as a very demanding customer and hence many firms have used CERN as a reference for the quality of their products. This helped, for instance, a firm producing cooling equipment and another firm fabricating light guides and scintillators for solar energy to extend their markets. In some cases a CERN contract stimulated the collaboration between firms with the consequence that the products of smaller firms became integrated into the products of large enterprises, with an ensuing increase of turnover and exports. This happened, for example, to firms producing hydraulic equipment.

These are just a few arbitrarily chosen examples. The spin-off from CERN procurement contracts was systematically and quantitatively investigated in several studies. The first one [9, 10] was carried out for the period 1955–1973 and concerned mainly contracts for the Intersecting Storage Rings (ISR) and the Super Proton Synchrotron (SPS). This study was complemented by a second one [11, 12] extending up to 1985. About 6,000 firms were suppliers of CERN, of which 519 were identified as being involved in high-tech products. Of these, 160 firms were selected and members of the higher management were interviewed. During the interviews they were asked to assess the current and future economic benefits resulting from the CERN contracts. The results were scaled up for the 519 firms, representing a total value of purchases of about CHF 7 billion.

Finally a third study [13] (see also [14, 15]) covered the period 1997–2001 and concerned mainly LHC contracts. During the construction of the LHC, technology transfer was further extended, in particular in fields of superconductivity and cryotechnology. The protons will be kept on track by more than 1,000 superconducting magnets with a special design providing fields up to 7 T. To achieve such high fields, the magnets have to be cooled with superfluid helium at a temperature of 1.9 K. The construction of the magnets, the production of the superfluid helium and its distribution over distances of several kilometres presented completely new technical challenges.

This time, 629 technology-intensive companies were interviewed, with the following results:

- In 38% new products had been developed.
- In 42% their international exposure had increased.
- In 44% there was significant technical learning.
- In 17% a new market had been opened.
- In 14% a new business unit had been established.
- In 13% a new R&D unit had been started.

The economic benefit for a firm resulting from the execution of a CERN contract may be defined as the increase of turnover plus the saving in production cost for activities independent of the CERN contract. Thus, an economic utility can be

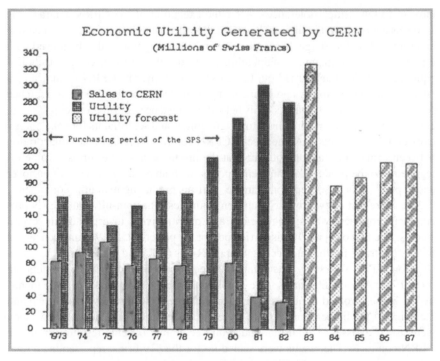

**Fig. 9.2** Economic utility created by CERN contracts defined as the increase of turnover divided by the value of the CERN contract. Information obtained from 160 firms in the period 1973–1987. The average economic utility was around 3

defined as follows: Economic utility = (increase of turnover + saving of production cost)/(value of CERN contract). In Fig. 9.2 the economic utilities (actual and predicted) for the firms and the value of the CERN contracts are displayed for the period 1973–1987.

Of course, the benefits will be different for various fields of technology as is shown in Fig. 9.3 and was found to lie between 1.4 and 4.2, with an average around 3.

The three analyses taken together indicate that for every euro spent in a high-tech contract a company will receive about three euros in the form of increased turnover or savings. This implies very crudely that in a laboratory such as CERN about one quarter of the budget is spent on high-tech products and consequently about three quarters of the overall public spending is eventually returned to industry.

## 9.2.4 Technology Transfer by People

My experience is that the most effective transfer of knowledge can be achieved through people. This can be done in many different ways, but it would take too

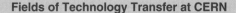
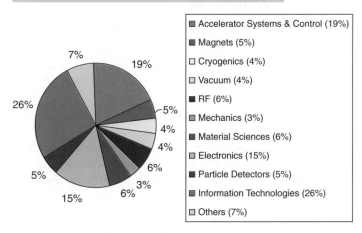

**Fig. 9.3** Distribution of technology transfer among various fields

much space to deal with this issue here in detail. One possibility is a transfer of experts between the laboratory and commercial firms, be it that people leave for good or that they are seconded for certain periods. In spite of its efficiency, this method was not used extensively and was restricted to only a few special cases, mainly in connection with procurement contracts. The reasons are complicated and cannot be analysed in depth here. CERN staff preferred, in general, the challenging and interesting environment in the laboratory compared with the more rigid tasks to be dealt with in industry. Industrial employees on the other hand were afraid to leave their firms for long periods since they might be excluded from promotion cycles or lose out financially otherwise. Firms sometimes expressed the opinion that employees would be spoiled in a laboratory without the strict frame for immediate return and hence were reluctant to send employees to CERN.

Expertise acquired at CERN can be useful in completely different areas. For example, a physicist busy at CERN trying to identify a few Higgs particle events among the billions of of background events by statistical (Monte Carlo) methods was hired by a bank, where he used the same method with minor changes to single out particular clients of the bank, e.g. those who were in danger of leaving.

The technology transfer through people works excellently when the training and the development of young researchers is included. A large fraction of the students involved in CERN experiments become competent with many of the technologies used in their work or are taught about them in many courses and they take this knowledge with them when they later go to industry. About 40% of the students working at CERN eventually go to industry. This is illustrated by an experience with one of my students. After he had been awarded a PhD degree, he was hired by a motor car firm to his surprise since there were many candidates. He was so astonished that he asked the personnel manager of the firm why he had been selected, with a PhD thesis in elementary particle physics and having had nothing to do with motor cars in his whole life. The answer was as follows:

> You have learned methods to deal with complicated problems, you have learned to work in a team and an international environment, you got accustomed to a well defined time and budgetary schedule, you are familiar with computing, simulation and networking in a practical way and you know how to handle electronics, magnets, vacuum and other useful techniques. This is more important than special knowledge which becomes obsolete within a short time.

Of course, working at CERN for some time implies that students become fluent in English and French, often decisive for their future jobs.

In conclusion, one can state that the technological spin-off from CERN and, in general, from elementary particle physics to industry is considerable, and may be even better than what has been achieved in some special technology transfer programmes organized by governments.

# References

1. Schopper H (1994) Was haben Quarks mit Computerchips zu tun? Wissensch Wirtsch Nov:35
2. Schopper H (1996) Bedeutung und Perspektiven der physikalischen Forschung, Festschrift Zwanzig Jahre Physikzentrum Bad Honnef, German Physical Society 1996
3. Schopper H (2004) Talk at the Forum Scienza Societá, at Genoa, Italy, 25 October 2004
4. Bell JS (1987) Speakable and unspeakable in quantum mechanics. Cambridge University Press, Cambridge
5. Aspect A et al (1981) Experimental tests of realistic local theories via Bell's theorem. Phys Rev Lett 47:460
6. Santos E. (2005) Bell's theorem and the experiments: increasing empirical support to local realism. Available via arXiv. http://arxiv.org/abs/quant-ph/0410193
7. Santos E. (2005) Bell's theorem and the experiments: increasing empirical support to local realism. Stud Hist Philos Mod Phys 36:544–565
8. Amaldi U (1999) In: International Europhysics conference – High Energy Physics '99, Tampere, Finland, July 1999
9. Schmied H (1975) A study of economic utility resulting from CERN contracts. CERN report 75-5. CERN, Geneva
10. Schmied H (1977)A study of economic utility resulting from from CERN contracts. IEEE Trans Eng Manage 24(4):125
11. Bianchi-Streit M, Blackburn N, Budde R, Reitz H, Sagnell B, Schmied H, Schorr B (1984) The economic utility resulting from CERN contracts (second study). CERN report 84-14. CERN, Geneva
12. Bianchi-Streit M, Blackburn N, Budde R, Reitz H, Sagnell B, Schmied H, Schorr B (1985) The economic utility resulting from CERN contracts (second study). CERN report 85-04. CERN, Geneva
13. Autio E, Bianchi-Streit M, Hameri AP (2003) Technology transfer and technological learning through CERN's procurement activity. CERN report CERN-2003-005. CERN, Geneva
14. Bressan B (2004) A study of the research and development benefits to society resulting from an international research centre. Dissertation, Faculty of Science University of Helsinki. Available via http://ethesis.helsinki.fi
15. Haehnle M (1998) R&D collaborations between CERN and industrial companies, Wien Wirtschaftsuniversität, IIR-Discussion 56

# Chapter 10
# Unloved but Necessary – Management and Finances

The scientific and technical excellence of CERN has never been in doubt. However, the CERN member states were wondering whether the management, administration and the personnel policy could not be improved. The purpose was, of course, to reduce their yearly contributions without damaging the scientific programme.

## 10.1 The Kendrew Committee

The first blow came from the 'Kendrew Committee', which had been set up in March 1984 by the Advisory Board for the Research Councils (ABRC) and the Science and Engineering Research Council (SERC) to review the engagement of the UK in particle physics, in general, and its financial contribution to CERN, in particular. The committee was chaired by Sir John Kendrew, a famous biologist (Nobel Prize 1962) and former director of the European Molecular Biology Laboratory (EMBL) in Heidelberg. I got to know Kendrew quite well when I was director of DESY in Hamburg during the 1970s. Together we created at the electron storage ring DORIS the first outstation of EMBL using synchrotron radiation for biological investigations. He was a very kind and objective person, a real gentleman, passing away much too early in 1997.

After more than 1 year of deliberations the report [1] of his committee was published and a magazine [2] (see also [3]) wrote, "Kendrew takes knife to high-energy physics". The committee reckoned that CERN is "the leading particle physics laboratory in the world" and that "British withdrawal from CERN would be a major blow to the science both of the UK and the rest of Europe, and it would have long-term detrimental implications for international collaboration", but nevertheless costs must be cut, in spite of the fact that the CERN total budget had already decreased by 20% in real terms over the previous decade.

The main recommendation was that the UK should remain a member state of CERN until 1989, when the construction of the first phase of LEP would be completed, but it should only continue if the other member states agreed to cut the UK subscription by 25% by 1990–1991. Domestic expenditure should also be reduced by the same percentage. The UK contribution to CERN amounted to about 16% at that time and it was a futile hope that the other member states would step in and take

H. Schopper, *LEP – The Lord of the Collider Rings at CERN 1980–2000*,
DOI 10.1007/978-3-540-89301-1_10, © Springer-Verlag Berlin Heidelberg 2009

over part of the UK contribution. Hence, CERN seemed to face a serious crisis at a crucial moment when the LEP construction was in full swing.

The Kendrew report dismayed, of course, the British particle physicists and it was denounced by many of them, with Christopher Llewellyn-Smith of Oxford University taking the lead. Criticism from all over Europe followed. It took me some time to find out why the UK was in such a particularly difficult situation. The contribution to CERN had to be financed from the budget of SERC, which also had to cover research activities at national universities. This created two problems, the second of which also produced negative effects in other countries.

The CERN budget has to cover all expenses – operation, investment and personnel. Of these, only a relative small fraction goes directly into research. On the other hand, in most countries the real research activities at universities were financed by special organizations such as SERC in the UK, INFN in Italy, DFG in Germany and CNRS in France, whereas the main costs of the infrastructure (e.g. buildings, heating, administration) are borne by the universities. Compared with the budgets of the research organizations, the CERN budget appears large since it includes all the infrastructure costs. Hence, it should rather be compared with the total budget of a university. This then reveals that the CERN yearly expenditures correspond to that of a single large European university. The member states of CERN finance several hundred universities, and CERN could be considered as just one more. The wrong perception that the CERN budget is exceptionally high stems therefore from the wrong comparison of two budgetary figures, one containing only research funds, the other covering also infrastructure and personnel.

Because of this difference in allocating budgets, the UK CERN contribution amounted to about 20% of the total SERC budget mainly destined to finance research but not infrastructure and personnel. Such a problem did not arise in Germany, where the CERN contribution is contained in a federal budget line including other large international organizations, but not university budgets.

The second problem was related to the condition that CERN contributions have to be paid in Swiss francs. Any change in the conversion rate had to be compensated for within the budget of SERC. This did not make any sense since there was no reason why the national UK activities should be reduced because the value of the Swiss franc went up relative to that of the pound. In most other countries the yearly contributions to international organizations such as CERN, EMBL and the European Space Agency are largely decoupled from the national budgets such that currency fluctuations do not affect the national programmes.

When Prime Minister Margaret Thatcher visited CERN I explained these difficulties to her. At the end of her visit she told journalists that she had been convinced that the contribution to CERN is well spent and she promised to look into the problem. Indeed I learned later that she provided some extra funds to SERC to compensate for the changes in the CERN contribution. A British magazine published a cartoon showing the chairman of SERC with a hat collecting money poured into it by Mrs. Thatcher. However, the hat had a hole through which some of the money fell out and I was sitting below the hat collecting it. Nevertheless, SERC insisted that it was its decision how to distribute the total funds allocated to it and apparently it did not use the additional money as compensation for the CERN problem, but used it for other purposes. The UK pressure continued until 1987, when the UK government decided not to give notice of withdrawal from CERN, a decision relayed to parliament by the higher-education minister, R. Jackson, under the condition that major savings could be made following recommendations of another committee which had been set up by the CERN Council in the meantime.

## 10.2 The Abragam Committee

At the instigation of the UK, the CERN Council appointed[1] an ad hoc consultative group of external representatives on 20 February 1986 with the task of conducting an in-depth comprehensive review of CERN investigating human and material resources, employment conditions, cost-effectiveness and finding new sources of funding. Tacitly the member states expected suggestions for how the budget of CERN could be reduced by reviewing its management structure, staff complement and pay and working conditions. Again the scientific and technical activities were not in doubt and were not to be evaluated. A report with the findings and recommendations was expected within 1 year. Of course, this meant additional efforts for the CERN management, already stressed by the construction of LEP.

This review committee was chaired by Anatole Abragam (Fig. 10.1), a very interesting and fascinating person whom I got to know during the activities of the committee, which offered many occasions for private discussions. Born in the Soviet Union, he grew up in a peaceful family, his father being an industrialist fabricating buttons and his mother a physician. The Jewish Abragam family left Moscow in 1925 when anti-Semitism grew in the Soviet Union. Anatole went to Paris, where he studied Greek, left his medical studies after 18 months and collected all kinds of diplomas. Finally in 1946 he met Francis Perrin, Head of the French Commission of Atomic Energy (CEA) and a famous physicist, who gave him a non-permanent job at CEA. However, being an independent person and a free spirit, he did not feel at home in an organization which was organized according to military principles. He went to Oxford, where he received a doctorate but refused a chair in theoretical physics. In 1965 he went back to CEA and became the director of the physics department, a post he held until 1971. In spite of his disgust for big organizations he carried out this job with great efficiency. However, he found his true destination at the Collège de France, where he was appointed to the chair for nuclear magnetism, a field in which he excels. Here he could interact with few people, exchange independent thoughts and develop his qualities in an academic environment. He was elected to the Académie Française des Sciences and he received many distinctions. But, as I can guess from many conversations, he cannot understand, with some bitterness, why he was not awarded the Nobel Prize. He is a man who is true to his principles, fair and outspoken, very critical, sometimes with some sharpness and acidity. His personal preference for small physics and a sceptical attitude towards big science was well expressed when at the end of the work of his committee he gave me a

---

[1] CERN/1609/Rev.2, 20 February 1980

**Fig. 10.1** Anatole Abragam, a renowned French physicist, chairman of the CERN review committee

copy of his book *Reflections of a Physicist* [4] containing the chapter 'Big Science Versus Little Science' with the somewhat sarcastic dedication *"To Herwig Schopper who thrives to make 'Big Science' 'Great science' "*. He was now in charge of the evaluation which was supposed to create a new frame for the future of CERN.

The seven-member committee included two industrial magnates, Carlo de Benedetti, managing director of Italy's Olivetti, and Haakon Sandvold, director of Norway's Ardal og Sunndal Verk, a Swiss management consultant, Jean Vodoz, president of AMYSA, and Miguel Boyer, president of Banco Exterior de Espana and former finance minister of Spain. Sciences were represented apart from by Abragam by Brian F. Fender, a physical chemist, Vice-Chancellor of the University of Keele and former head of ILL in Grenoble, and Wolfgang Paul, a physicist from the University of Bonn and later Nobel Prize winner. The committee had two advisers, Christopher Llewellyn-Smith, a theorist who later became director-general of CERN, and Pierre Petiau, a particle physicist working at the French Ministry of Research.

Most of the committee members did not know each other and at the beginning some funny incidents happened. In the first meeting Abragam had distributed specific tasks to the various committee members, yet nobody had been assigned to look at the finances. Since the Spanish member had not been given a task yet, Abragam asked him whether he had some experience with financial matters. The answer was: "Yes, some, I was Minister of Finances in Spain for several years." Having been minister, Boyer had the habit of checking into one of the most expensive hotels in Geneva and hired a limousine with a chauffeur. Since CERN had to pay for the expenses of the committee, I told Abragam that this was not acceptable. However, Abragam was afraid of telling this to Boyer and in the end, as usual, it was left to me as director-general to communicate unpleasant messages! It was not easy for some of the committee members to understand the unique structure of CERN. Indeed one of them remarked that an organization that is a complicated mixture of an industrial enterprise, an academic institution and an international organization could not be handled at all.

The official mandate of the committee was

> ...to advise CERN Council how human and material resources, employment conditions, structure, operation and future use and development of facilities might be developed to operate with maximum cost effectiveness and value for money at alternative levels of funding by present Member States, and to assess their consequences for the CERN programmes and for the services to the Member States.

and to "assess the possibilities for engaging and enlarging other sources of funds and resources." In a footnote, the UK delegation requested the consideration of a budget level reduced by 25%. The committee met 14 times and interviewed a large number of people, from both inside and outside CERN. This was certainly an additional load not only for the management but also for many CERN staff members in the already critical situation of realizing LEP. Since the mandate of the committee was so wide, it was not quite clear what to expect from its recommendations and some feared the worst. The first recommendation was quite positive:

> The roots of CERN's excellent scientific record lie in the supranational scientific enthusiasm which prevails there. . ... This enthusiasm is directly ascribable to the world leadership currently enjoyed by CERN in its field of research. . ... If CERN were to lose this leadership and ceased to be the focus of excellence, it would lose its main *raison d'être*, its attractiveness and its dynamic qualities.

The committee declined to consider lower funding levels and alternative scientific programmes. "The committee considers that an *a priori* reduction of the budget will inevitably jeopardize CERN without giving a rationale for future management practice. A reduction of the program is in any case premature as a conclusion." This referred without any doubt to the proposed upgrading of LEP to phase 2 which was just under discussion. The committee complimented CERN on the building of LEP without increasing the existing staff, but complained at the same time that the services to users had deteriorated.

After these positive statements the committee stated, "However, . . .excellence in the scientific field alone is no longer enough. It must go hand in hand with excellence of management, in the use of resources and in the services offered to users. In the latter respects, CERN has fallen behind and must catch up systematically and quickly." The committee confined itself to suggesting the following management measures:

- An early-retirement and early-departure programme for between 330 and 500 staff in 1988 and 1989
- No further granting of international status to non-professional and technical staff members, so that their social benefits would be borne by the Swiss and French social security systems

- A new personnel policy stressing assessment of performance and promotion based on merit
- In the near future minimum recruitment of new staff and minimum granting of indefinite contracts,
- Several suggestions concerning the pension fund and financial accounting

One major push of the recommendations was to create *more mobility*. De Benedetti, who was described as a tycoon who had refurbished and saved Olivetti, asked me how many people left CERN every year. When I replied about 5–6%, his reaction was, "Then CERN is a completely sclerotic organization." When I asked what the corresponding percentage was at Olivetti, he said about 30%. When I inquired how many had been fired because of bad performance he became embarrassed and had to admit "nobody". It turned out that the mobility at Olivetti was achieved by an early-retirement programme which according to Italian law was financed almost completely by the Italian state. When the committee decided to suggest to Council an early-retirement programme, I insisted it should also propose that the compensation to the capitalized CERN pension fund be provided as an extra contribution. The committee did propose this, since in private discussions it admitted that if CERN were to save money in the long run, it would be necessary to invest money in the short run for severance pay, to compensate for losses from the pension fund and for retraining. However, when Council decided to introduce such a programme it was done without any compensation for losses from the pension fund and Council simply said, "You have to find the money"!

The emphasis of the recommendations of the Abragam committee was on personnel policy and I shall come back to some of the issues. The general outcome of this review did not change CERN fundamentally. When I expressed my doubts about the usefulness of external reviews to an expert from a management-consulting firm, he admitted that such reviews produce very little new information of which the management is not already aware. He claimed that the main benefit of external reviews is that they provide the management with a tool to convince the staff and/or the superior authorities that some painful measures need to be carried out. To a limited degree the recommendations of the Abragam committee indeed turned out to be useful in this respect. However, they did not result in the hoped for savings of 25%. Nevertheless, in the end the UK could be convinced not to quit CERN.

The organization as it existed with minor modifications during the construction of LEP is shown in the Appendix.

## 10.3 Personnel Policy

For a long time the personnel policy of CERN had been and still is today a critical issue. The reason is that the CERN staff expect a salary scale comparable to that of other intergovernmental organizations. Indeed, regular comparisons showed that CERN staff receive salaries which are about 15–20% lower than those of staff of other intergovernmental organizations, e.g. the European Space Agency. On the other hand, CERN outside users compare the CERN salaries with their academic salaries at home and find that these are considerably lower, in particular when taking into account that CERN remunerations are tax-free in Switzerland and France. The result is a battle which has been going on a long time between Council and the staff, with the director-general sandwiched in-between.

This concerns in particular the yearly index of inflation. Council has adopted a complicated formula for calculating this index by taking various averages of some of the costs in member states and local organizations. A separate index is calculated for the material budget and the personnel cost. Council was prepared to accord in most years, although not always, the full compensation for the material index, respecting partially in this way the gentlemen's agreement that LEP would be built with a constant budget (see Sect. 3.4). On the other hand, it very often reduced the index for the personnel cost. This made the Staff Association very angry since it thought that the calculated index should provide an automatic compensation, whereas Council considered it only as a guideline. Indeed, in some countries any automaticity in compensating for inflation is in principle refused since such a mechanism is considered to drive inflation. For such reasons Council insisted on preserving its complete freedom in deciding each year on the budget, in general, and on the index, in particular. The Staff Association, however, was prepared to accept any policy, even a disadvantageous one, under the condition it avoided any 'arbitrariness'. However, the main policy of Council was to establish no policy in this respect. As director-general one has a hard time to explain these different views to both sides.

The age distribution of the staff certainly presented a problem, as was recognized rightly by the Abragam committee. Like many other laboratories, CERN saw a rapid expansion in the golden years 1950–1970, when the number of posts was steadily increased. When this development ended, all available posts were occupied and with almost no recruitment the average age increased every year by almost 1 year. In 1986 it had reached 45.5 years. Thus, it was reasonable to introduce a departure programme to rejuvenate the staff, and not only to reduce the personnel expenditure. In long discussions we managed to convince Council that about one third of the vacancies created by early departures could be filled by young people.

In my opinion the departure programme, in principle justified, was exaggerated by Council by even continuing it until recently (see Fig. 10.2). The first target was to reduce the staff complement to about 2,500, which during the construction of the LHC was lowered further to about 2,000 . In order not to damage the programmes, it was suggested to outsource many activities, a fashion which is also considered to be modern management in industry. If the outsourcing concerns routine services such as cleaning rooms or distributing mail, this may be a reasonable policy; however, if it is also applied to core activities such as some engineering work or computing and networking activities, there might be nominal savings but at the cost of efficiency and loss of competence. When I discussed this much later with Council delegates, they privately admitted that they might have gone too far. In fact, the early-departure programme did not help to reduce the costs for personnel and hence also the total budget during the LEP construction period stayed more or less constant (see Fig. 10.3) because of the additional expenses for the pension scheme and for the outsourcing. Indeed, to avoid raising the personnel budget we had to introduce a new personnel management policy – 'personnel management by

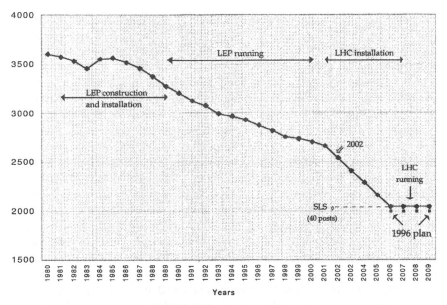

**Fig. 10.2** The development of the CERN staff complement with time

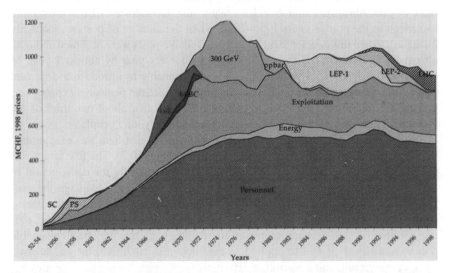

**Fig. 10.3** The total budget of CERN. After a fast rise in the early years and specific funds for the Super Proton Synchrotron (SPS) machine (300 GeV), the budget dropped and stayed constant after 1981 during the construction of LEP. The personnel budget amounts to more than half the total budget

budget' and not according to the requirements. When a decade later the LHC ran into financial problems, the reduction of staff had to be extended to its extreme and only then were appreciable savings possible, with the danger that CERN staff would lose competence.

For the construction of LEP the main personnel problem was to activate the necessary personnel inside CERN. No additional posts were approved and it was natural that in an organization which had existed for 25 years strong human bonds had been created in existing divisions and departments, not to speak of the two separate laboratories CERN I and CERN II. To break the borders between existing units and transfer personnel to new units adapted for the realization of LEP was a major management problem. The appointment of Emilio Picasso as LEP project leader was one important element as mentioned already (Chap. 3). In the end it was the common efforts of the directors and division leaders, some of them newly appointed, that made it possible to achieve a remarkable internal mobility. This was not sufficiently appreciated by the Abragam committee, but at least the then president of the Council, Wolfgang Kummer, felt [5] "that Schopper achieved something unbelievable in shuffling 1,000 of the 3,500 staff members to build LEP", and he believed that "compared to other international organizations CERN management is doing very, very well", especially considering that other international organizations operate within "much more comfortable budgets".

## 10.4  The LEP Management and Budget

Since LEP had to be realized with the existing personnel and financial resources, it was not possible to set up an autonomous LEP unit with the appropriate allocations of resources. Since LEP was using the existing CERN accelerators as injectors, it became fully integrated into the basic programme of CERN. Hence, the LEP activities were so closely interwoven with the rest of CERN that LEP had to benefit directly or indirectly from almost all parts of CERN. Because of this situation the management of LEP had to be rather unconventional.

A LEP division was established which had the main task of taking care of the main ring of LEP. With some effort I could convince Günter Plass, who had been responsible for many years for the Proton Synchrotron (PS), to become division leader. To coordinate the various activities spread all over CERN, a LEP Management Board was established with Emilio Picasso as chairman. Among its members one could find (responsibilities in parentheses) Giorgio Brianti (director responsible for the accelerators, except LEP), Wolfgang Schnell (rf system), Lorenzo Resegotti (magnets), Hans-Peter Reinhard (vacuum), Henry Laporte (building), Manfred Buehler-Broglin (finances) and Günther Plass (LEP division leader). Group leaders were invited depending on the agenda, e.g. Gerard Bachy (installation) or Bas de Raad (injection). Picasso, with his charming Italian temperament, managed to guide the sometimes heated discussions into calm waters and in the end all decisions were taken on the basis of objective arguments or reasonable compromises.

I attended most of the meetings as a silent observer. This allowed me to understand any upcoming problems in detail and when requests had to be decided in the Directorate I was well informed about the priorities. Sometimes, but rarely, I intervened to stop at the source developments which I considered to be against general CERN policy or even dangerous. For example, I insisted that the contract for the electronic control system of LEP should be awarded to a commercial firm which had great experience. However, later it turned out that the firm lost interest in the CERN contract since it had hoped to obtain a similar order from the military, which would have made the whole business profitable. It cancelled the contract, which was critical for the construction of LEP, and an enormous effort had to be made by the CERN experts to solve the problem and keep the construction of LEP on schedule. Directors-general are not always right!

The way in which we handled the LEP budget was completely unorthodox and against all modern management rules. Since the available funds were extremely small we decided not to allocate definite sums for the various components since we were sure that the allocated money would certainly be spent. Hence, we asked the various group leaders to do their best and carry out their task with as little funding as possible. Of course, Picasso and I had prepared a secret budget plan, which we revealed to only very few people who were necessary to follow it up. In the end it worked, perhaps even to our surprise, thanks to the ingenuity, competence and conscientiousness of the group leaders. The surprising result was that the machine parts, which themselves consisted of high-tech components and involved a considerable technical risk (e.g. magnets, vacuum, rf), were cheaper than estimated, whereas the infrastructure (e.g. cooling, ventilation), which had to be bought to a large extent off the shelf, turned out to be more expensive.

This is illustrated in Table 10.1. The second column shows our original 'secret' estimates. The third column gives the actual cost in 1981 prices. Mainly because of the increase of the cost of conventional material we had to ask for a contingency which was not accepted when the project was approved ('time is contingency'). The last column shows the prices in 1986 scaled by the inflation indices as approved by

**Table 10.1** The LEP budget

|  | Cost (millions of Swiss francs) | | |
|---|---|---|---|
|  | Estimation 1981 | Actual cost 1981 prices | Revised cost 1986 prices |
| Machine components (e.g. magnets, vacuum, rf) | 335 | 294 | 353 |
| Machine infrastructure (e.g. cooling, ventilation) | 120 | 208 | 250 |
| Injection system | 80 | 92 | 111 |
| Tunnelling | 310 | 280 | 364 |
| Surface buildings | 45 | 43 | 54 |
| Intermediate total | 890 | 917 | 1,132 |
| Contingency | 'Time' | 126 | 162 |
| Final total | 890 | 1,043 | 1,294 |

Council. These 1986 revised cost estimates take into account the additional cost due to the difficulties of the tunnelling work after approval by the Finance Committee. In 1990, 6 months after the commissioning of the collider, the final balance sheet of LEP could be drawn up. The Finance Committee stated in the final LEP financial report:[2]

> The final cost of 1300 MCHF at current prices is to be compared with the 1294 MCHF authorized by the Finance Committee in 1986. This result is due to the ingenuity of the designers and to the excellent standard of work of the contractors and CERN staff. As a result of the Member States' support, Europe has an experimental complex of first-rate world importance for the 1990s with considerable potential for further development'.

The full implication of this statement became obvious when LEP was upgraded to phase 2 with 105-GeV beam energy and its 'final phase', the LHC.

## 10.5  The Total CERN Budget

Looking at the evolution of the total CERN budget (see Fig. 10.3), one notices a dramatic change around the period 1973–1980. After a rapid increase in the 'golden years' up to 1975, a large peak can be seen which is associated with the financing of the Super Proton Synchrotron (SPS; 300 GeV), the last project at CERN for which additional funds had been approved. After the completion of the SPS, the budget drops and from 1980 on is more or less constant. With this reduced constant budget LEP had to be built and later the LHC. The small increases are due to the joining of Spain and Portugal. When these two countries became member states, I had to face tough negotiations since the old member states wanted to use the contributions of the newcomers to reduce their own contributions. This was not justified since with the new countries more users and other additional commitments were to be expected. In the end a compromise was achieved and at least some of the new money (about CHF 55 million per year, corresponding to about 7% of the total budget) could be added to the CERN budget, certainly a welcome release.

With the total budget level and the personnel budget being almost constant, the construction money for LEP had to be obtained from the existing material budget of about CHF 400 million per year. In the most critical years 1986–1987 about half of it had to be used for construction of LEP, implying hard sacrifices for the ongoing scientific programmes (see Fig. 10.4). Even making all possible savings, some indispensable expenses remained, e.g. electricity and water (CHF 50 million) or site operation (buildings, computing, insurance, postal and telephone costs, together CHF 70 million), leaving only about CHF 80 million for the exploitation and refurbishing of all the remaining scientific programmes, above all the proton–antiproton experiments at the SPS which had led to the award of the Nobel Prize.

A more detailed but also quite annoying issue concerned the cash flow. The natural profile of the yearly cash requirements for any large technical project resembles a

---

[2] CERN/FC/3313, 30 May 1990

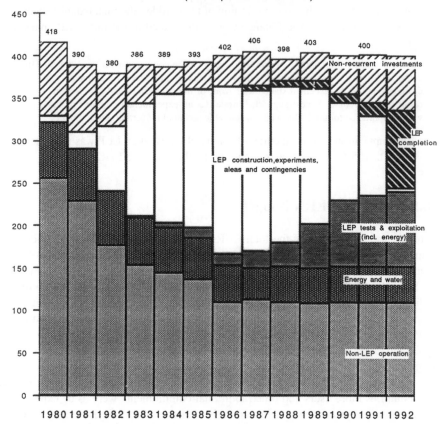

**Fig. 10.4** The material budget of CERN. The 'bubble' for the construction of LEP 1 had to be cut out of the existing budget. *LEP completion* stands for the upgrading to LEP 2

bell-shaped curve. The expenses increase only slowly in the first years of the project, reach a maximum at about half way through when most of the orders to industry have to be paid and finally tail off. To accommodate such a cash-flow profile into a very restricted constant overall budget is extremely difficult. Of course, one can try to advance some payments by making down payments, but delaying some payments until after the completion of the project is unavoidable, i.e. creating debts. There are limits to such an operation and as a result the construction of LEP had to be extended by about 1 year owing to the limitations of funding. In Fig. 10.5 the necessary payment profile is compared with the available funds. As one can see, the limited budget resources made it necessary to postpone some of the payment to the years after 1989, which implied creating debts. The accumulated savings at the beginning of the LEP construction and the debts towards the end are displayed in Fig. 10.6 and

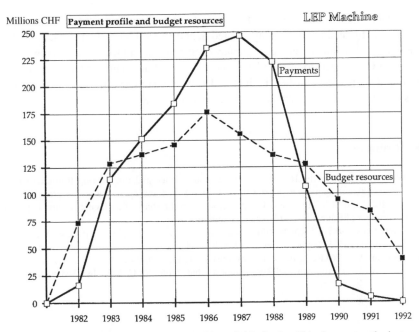

**Fig. 10.5** Payment profile for LEP compared with available funds within the constant budget

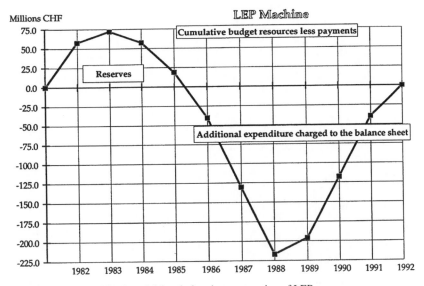

**Fig. 10.6** Accumulated funds and debts during the construction of LEP

they reached a maximum at about the time I left as director-general. Of course, my successors blamed me for having created debts! When I had asked Council to allow bank loans to be taken out, such a request was refused and we had to borrow money from the CERN pension fund. When the LHC was built, again the resources had to be found within the constant budget and naturally similar difficulties arose, but in this case Council agreed at least to take out bank loans.[3]

# References

1. Advisory Board for the Research Councils and Science and Engineering Research Council (1985) High energy particle physics in the United Kingdom: report of a review group set up by the ABRC and SERC to review UK participation in the study of high energy particle physics. Department of Education and Science, Stanmore
2. Anonymous (1985) CERN: Kendrew takes knife to high-energy physics. Nature 315:619
3. Cambel P (1985) Schopper's views on Kendrew. Nature, 315:708
4. Abragam A (1986) Reflections of a physicist. Clarendon, Oxford
5. Sweet W (1987) Abragam and Rubbia reports chart future for CERN. Phys Today 40(9):71

---

[3] In the end the situation for the LHC became more complicated since mainly because of the very advanced and complicated cryogenic magnets and the cooling system the total cost had to be increased and Council provided some compensation for it.

# Chapter 11
# How To Invite the Pope? – VIP Visits

Because of its uniqueness and intellectual challenge, LEP became a kind of symbol for human endeavour to explore the unknown. Hence, it may not be surprising that it attracted as visitors many high representatives from politics, culture and even religion. The culmination was reached with a visit by Pope John Paul II on 15 June 1982. He was followed shortly thereafter by the Dalai Lama and several heads of state and other dignitaries came on various occasions. Some of these visits were not restricted to just formalities, but provided the opportunity for interesting discussions or sharing amusing stories. I would like to share some of them with the readers.

Often I was asked, "How does one invite the Pope, do you just write a letter to him?" Of course, it is not that simple. Among the international CERN staff one can always find someone who has special relations. Indeed, it turned out that a collaborator in the administration had served for several years in the Vatican and he explored unofficially the possibility of a visit. It turned out that the Pope intended to visit Switzerland; however, according to diplomatic protocol it is not seen with pleasure if a head of state visits an international organization on the occasion of a state visit to a country. Fortunately we learned that on another occasion the Pope wanted to pay a visit to the International Labour Organization (ILO) in Geneva, and this could be combined with a visit to CERN. After this had been clarified by informal channels, I was able to write an official letter inviting the Pope.

The style of CERN is traditionally one of pragmatism and familiarity, avoiding formal pomp as much as possible. We tried to apply this habit also to the visit of the Pope. This was facilitated by the fact that since the foundation of CERN Polish scientists had participated strongly in the CERN programme and hence it was also easy for several Polish scientists to be present and it turned out that one of them was even a schoolmate of the Pope. The Pope certainly enjoyed the unconventional atmosphere at CERN, so different from that to which he was accustomed. We showed him some of the installations and he gave a public address to the people present at CERN in beautiful summer sunshine (Fig. 11.1).

Fortunately there remained sufficient time for some serious discussions. I explained to him that in the collisions between matter and antimatter these are transformed into 'pure' energy and in a second step new kinds of particles are *created*. To demonstrate this process I used a diagram which I had invented a long time

ago at DESY showing colliding strawberries which are converted into heavier fruits such as bananas or pears (Fig. 11.2). Sometimes even a fish might appear, which could lead to the award of a Nobel Prize! When I showed this picture to John Paul II he protested and remarked, "You cannot *create* matter, creation is my business; you can only *produce* matter." Agreeing I crossed out under his eyes the word 'creation' and replaced it with 'production' (Fig. 11.2 shows the original diagram used on this occasion). At the time I considered this to be a funny incident, but its relevance became clear to me only much later. Indeed when I read scientific or popular magazines I find the word 'creation' in another context, namely 'creationism', which denies that evolution is compatible with science and in general terms evokes the relation between religion and science. In that respect the following dialogue with the Pope is interesting.

When I asked John Paul II what would happen if we find in science something which contradicts the teaching of the church, he answered with a quotation from Galileo Galilei saying that God had created two books, nature and the holy script, and hence a contradiction was not possible. When I insisted on a possible conflict, he retorted that in such a case the interpretation of the Bible would have to be changed. Such a liberal answer from the mouth of the Pope who on other occasions showed a rather conservative attitude surprised me. In the end I happily and full heartedly agreed to the statements which he made during his public speech[1] at CERN:

---

[1] Brochure *The Visit of the Pope to CERN* published by CERN and authorized by the Vatican on the occasion of the visit on 15 June 1982.

**Fig. 11.2** Strawberries and antistrawberries 'annihilate' each other and become 'pure' energy, which is subsequently converted into new fruits. At the request of the Pope, I changed 'creation' to 'production'!

There was a time when some scientists were tempted to take refuge in an attitude imbued with 'scientism', but that was a philosophical choice rather than a scientific attitude, as it tried to ignore other forms of knowledge; this tendency now seems to belong to the past. The majority of scientists admit that the natural sciences and the scientific method based on experiments, whose results can be repeated, cover only part of reality or rather reflect a particular aspect of it. Philosophy, art, religion and, above all, religion which is knowingly inspired by a transcendental revelation, perceive other aspects of the reality of the universe and above all of mankind.

Presently many discussions take place concerning the relations between science and religion. I believe many of the apparent conflicts could be avoided if the complementarity of the two domains were recognized.

Finally I asked the Pope what he thought about the case of Galileo Galilei. He explained that the condemnation of Galileo in 1633 had nothing to do with the conflict between science and faith. He said Galileo was a hard-headed Florentine who did not recognize the authority of the church and therefore had to be taught a lesson. When I asked why then he was not rehabilitated, the explanation was that no pope had signed a document against Galileo since his sentence was pronounced by the Holy Officium (at that time known as the Inquisition) and hence as Pope he could not change it. He would have to convince the cardinal[2] who was heading that office. Eventually John Paul II managed to convince the Vatican (even a pope has to obey certain rules!) and in 1992 Galileo was fully rehabilitated.

A few months after the visit of John Paul II, the Dalai Lama sent me a message asking to be invited to CERN. He wanted to discuss the similarities and differences between the vacuum in physics and the vacuity in Buddhism. Of course, I was pleased to organize such a meeting (Fig. 11.3), mainly with theorists including Leon van Hove, John Ellis, Maurice Jacob and John Bell. The outcome was surprising. It turned out that neither the vacuum in physics (see Chap. 8) nor the vacuity in Buddhism is really empty, but is rather filled by phenomena which provide a basis for a great unification. Unfortunately several hours was not sufficient to clarify all points. Privately I asked the Dalai Lama whether he thought that a disagreement between science and Buddhism could arise. His answer was that it could not since the two were complementary and in the case of a conflict the interpretation used in the teaching of Buddhism would have to be adapted. When I expressed my surprise that he gave the same answer as the Pope he said it was not surprising at all since the two often had dinner together and agreed on these issues. When I met the Dalai Lama again in 2008 at a meeting of Nobel Prize laureates in Petra (Jordan) I asked him whether he remembered his visit to CERN. He immediately answered: "Of course, you explained to me the quarks."

The 'entanglement' between science, religion and CERN continues. As a result of the recent start-up of the LHC in the LEP tunnel, this large project is attracting the attention of philosophers and other non-physicists [1].

A very enjoyable visit was that of Queen Beatrix of the Netherlands, who was accompanied by her husband Prince Claus (Fig. 11.4). Both were very interested in

---

[2] In 1982 it was Cardinal Ratzinger, later to become Pope Benedict XVI.

**Fig. 11.3** The Dalai Lama
and the author at CERN

learning what our objectives were. At lunch I asked her what her general impression of CERN was. She told me that she had visited CERN many years before when she was a young princess and that she was quite impressed by the development the laboratory had undergone in the meantime. She also related that from her first visit she had two recollections, a good one and a bad one. She agreed to tell me the bad one first. When the functioning of an accelerator (it was the PS) had been explained to her, she asked whether the speed of the particles increased all the time or whether there was a limit. The answer she got from the CERN guide was, "This is a stupid question." I apologized to her saying that the answer was stupid, but not her question, since according to Einstein's theory of special relativity the maximum speed of any particle cannot exceed the velocity of light. When I asked her about the good event, she said that at that time she used to drive her car herself and just in front of CERN she caused a little traffic accident. However, the CERN fire brigade (the *pompiers*) came immediately and to her relief the whole problem was settled within minutes.

One of the most interesting visits was that of Margaret Thatcher. When she came accompanied by her husband (Fig. 11.5), a nice, elder gentleman whom she treated with astonishing tenderness, she demanded not to be treated as prime minister but as a fellow scientist. This request was based on a degree in chemistry which she had been awarded by the University of Oxford. Indeed, she showed a remarkable interest and we had very frank and open discussions. The first question she asked was why we were going to build a round collider and not two linear colliders. A very pertinent question for electron–positron colliders as discussed in Chap. 2! Of

**Fig. 11.4** Queen Beatrix of
the Netherlands with her
husband Prince Claus

**Fig. 11.5** Prime Minister
Maragret Thatcher with her
husband Dennis (to her *left*)

course, she must have been briefed on this issue but I found it remarkable that she
took the time to get involved in such a technical issue. We managed to convince her
with the arguments given in Chap. 2 (see also Fig. 2.1). She then asked how big

**Fig. 11.6** Margaret Thatcher
involved in technical
discussions. From the *left*:
Fritz Ferger, Wolfgang
Schnell, Emilio Picasso,
Margaret Thatcher, author

the next ring to be built at CERN after LEP would be. When I replied that there would not be a bigger ring she retorted: "When I visited CERN many years ago when the SPS was under construction and asked your predecessor John Adams this question he gave me the same answer, which, as LEP proves, was obviously wrong. So why should I believe you." Figure 11.6 of course, I believe my answer was true and is still true today. To make a valuable step beyond LEP and the LHC, a new ring would have to be about 10 times as big, which is not realistic. I am convinced that there will be future projects at CERN, but not just involving increases of size but also involving new technologies or new ideas. CERN could be a location for the International Linear Collider (ILC) which was mentioned in Chap. 2.

At the time of Thatcher's visit the discovery of the W particle was imminent and that was mentioned to her. Before leaving CERN, she told me that she did not want to learn about such news from the press, but wanted to be informed privately. When Carlo Rubbia (UA1) and Pierre Darriulat (UA2) showed me some positive evidence for the W particle later that autumn, I was scared that something might leak out to the press since nothing can be kept secret at CERN. I therefore wrote a letter to Thatcher, outlining the preliminary evidence but asking her not to make it public. A few days later I received a reply assuring me of her strict confidence but asking again to be informed before we went public. In January 1983 I had to go to Japan for a meeting and before leaving I asked Rubbia and Darriulat to let me know before they wanted to publish their results. Indeed, when they informed me that they were ready for the announcement, I immediately sent a fax to Thatcher. When a few months later – just before the British general election – the Z particle was discovered, we again sent Thatcher the relevant documents. I was informed that she had taken these documents to her countryside retreat when she recovering from the election campaign.[3] If only more politicians took so much interest in science!

---

[3] I related this story in [2].

At the end of her visit she was interviewed by a few journalists, who also asked about the two evaluations mentioned in Chap. 10. She told them that she had become convinced that the money given to CERN was well spent.

There were a few occasions when the heads of states of the two host states, Switzerland and France, visited CERN. The most important one was the LEP ground-breaking ceremony on 13 September 1983 when President François Mitterrand of France and President Pierre Aubert of Switzerland acted as 'guest workers' by together putting the first stone into concrete (see Chap. 4).

Somewhat amusing were some events surrounding this ceremony which do not appear in any public reports. As part of the preparations for the ceremony I visited Aubert in Bern. He told me he would be pleased to attend the event, but knowing Mitterrand, he warned me that we would know only a few hours before the visit whether Mitterrand would really come. With that risk we prepared a programme foreseeing a ceremony in the morning, followed by a common lunch and in the afternoon a short tour through the laboratory. This proposal was accepted by the 'protocols' in Bern and Paris. However, the day before the event I received a call from Paris informing me that Mitterrand would have to leave immediately after the lunch because of other commitments and could not participate in the tour through the laboratory. Of course, I passed this news immediately on to Aubert, who replied that in such a case he would also cancel his visit to the installations. Another problem was the question of how the two presidents would arrive. The 'protocols' insisted that they had to arrive at the same time and be welcomed at the same mo-ment. It was decided that Mitterrand would come by helicopter, although at the time when CERN had been extended into France a special gate had been installed on the French side to be used particularly for such occasions. Waiting for the helicopter to arrive, Aubert was waiting in his car hidden behind a nearby building, whereas I was standing near the helicopter landing spot. After the helicopter had set down, we both came forward to shake hands with Mitterrand. Since I had warned Aubert about the limited time Mitterrand had, immediately after we got into a car Aubert anticipated Mitterrand saying he would have to leave early by remarking that he did not feel quite well and that he would unfortunately have to leave after lunch. Mitterrand expressed his satisfaction and agreement. But what happened?! During and after the ceremony both had the opportunity to talk to various people and they got to see a few installations. During lunch Jean Teillac, high commissioner of the French Atomic Energy Agency and president of the CERN Council, managed to convince Mitterrand to stay after lunch at least for a short tour through CERN. Although it was clear that Mitterrand was not in the best of health, he decided to stay. Suddenly Aubert felt better and the tour took place!

During lunch the two presidents discussed who the next director-general of UNESCO should be and one candidate was Abdus Salam, a theoretical physicist, Nobel Prize winner and founder of the International Centre for Theoretical Physics (ICTP)[4] at Trieste. Since I was sitting between the two presidents, they had to

---

[4] ICTP has devoted its activities to helping science in developing countries.

exchange their views rather loudly until I drew their attention to the fact that the person they were discussing was sitting a few metres from them and might be able to hear them. These are the pitfalls of getting involved with people of high political standing!

The occasion of installing the first magnet for LEP in the tunnel on 4 June 1987 was another occasion to welcome two representatives of the host states, Prime Minister Jacques Chirac of France and President Pierre Aubert of Switzerland, who had again been elected as president. In Fig. 6.13 they can be seen when installing the first magnet for LEP in the tunnel with the help of CERN experts.

Since there had been an attempt on Chirac's life, security measures were a major issue, as for all visits by VIPs. The close coordination between Swiss and French security services turned out to work perfectly and we never had a major problem. However, a few days before the visit of Chirac, I received a phone call from the French ambassador asking what means of transport was foreseen for the prime minister. When I answered, "by car", I was asked whether it would be an armoured car. When I replied that CERN had no such cars, I was informed that such protection seemed inevitable. I knew that the government of Geneva had an armoured car and when we asked for that car to be made available to CERN for a day, the answer was positive and it was even agreed for us to put a CERN number plate on it to make it neutral. Gladly I reported to Paris that the problem had been solved – uttering a sigh of relief. Too early! A day before the visit I received a call inquiring what make the car was. When the answer was "a Mercedes", I was informed that it was not acceptable to see the French prime minister on TV getting out of a car of non-French production. The only alternative I could offer was my official car, a rather old Peugeot 600, definitely not armoured. This was the final choice and the visit went ahead in full harmony.

Our general tendency was to make all the visits of high-ranking people as informal as possible and in most cases we managed to avoid too strict a 'protocol'. This was also true for the visit of the Spanish King, Juan Carlos I, who attended the meeting on the occasion of CERN's 30th anniversary (Fig. 11.7) with his wife and two children, two charming young ladies. Other examples of important visitors were Mario Soares, president of Portugal, and Gro Harlem-Brundtland, prime minister of Norway. When Richard von Weizsäcker, president of Germany, came, he made the ironic remark that we had invited the wrong one of two brothers. But I could assure him that the other brother, the well-known physicist Carl-Friedrich von Weizsäcker, was a regular visitor to CERN and we had an interesting exchange of views about the importance of science.

What is the purpose of such visits? My experience was that many of the politicians had completely wrong ideas about CERN. From the media reports they got the impression that the activities at CERN were rather abstract and theoretical, hard to understand and far from everyday life. However, seeing the installations at CERN, the technologies involved and meeting the many young enthusiastic people working here in an international environment resulted in such wrong impressions being corrected in most cases. Most politicians could even be convinced that the fundamental questions we are investigating are relevant for human society.

**Fig. 11.7** Ceremony at the 30th anniversary of CERN. Speaker Nobel Prize laureate Isidor Rabi, sitting from *left* Pierre Aubert (Swiss Federal Councillor), Sir Alec Merrison (president of the CERN Council), King Juan Carlos I, Herwig Schopper (director-general), Hubert Curien (French minister of research and technology), P. Brooks (UK under-secretary of state for education and science)

But LEP served not only as a catalyst for relations with politicians but also for relations with poets. One day Friedrich Dürrenmatt, the famous Swiss writer, visited LEP (Fig. 11.8). Of course, I complained to him that his play 'The Physicists' gave a completely wrong impression about how progress occurs in physics. Since the play is written around a theoretical physicist who speculates alone in his chamber about the riddles of the world and solves them by sheer reasoning, the importance of experimental observations is completely neglected. Dürrenmatt answered that he was well aware of the necessary symbiosis of theory and experiment for scientific progress since he had a physicist as a neighbour who had explained all this to him. However, in his play he only wanted to demonstrate that even the most intelligent people could be misused by a clever dictatorial regime. We also had an interesting debate concerning what 'understanding' means. He claimed that it is a difficult problem and told me the following story. A blind man was asked whether he knew what 'white' was. When he gave a negative answer, his arm was taken and bent into the form of a swan's neck and head. He was requested to explore the shape of his arm by touching it and finally was told, "Now you have an idea what a swan is and hence you know the colour 'white'." After his visit, Dürrenmatt sent me a copy of his book *About the Observation of the Observer of the Observers* [3],[5] a title relating to our discussion although the book is an exiting criminal story.

Dürrenmatt was accompanied by his wife Charlotte Keel, who was a filmmaker. She was so fascinated by the 'beauty' of the installations in the LEP tunnel, all the

---

[5] My translation. The original title is *Der Auftrag oder Vom Beobachten des Beobachters der Beobachter*.

**Fig. 11.8** The Swiss writer
Friedrich Dürrenmatt and his
wife Charlotte Keel signing
the CERN guest book

piping and the cables, that she came back later just to use it as background for a film that had nothing to do with physics.

Later Dürrenmatt called CERN an inverted NASA since there ever smaller particles are investigated by ever larger installations and he also said referring to CERN: "...it presents the most sensible what Europe has produced, because it is the apparently most senseless, associated to the speculative, to the adventurous, to pure curiosity."[6] In that context Wolfgang Frühwald [4], historian and former president of the Deutsche Forschungsgemeinschaft (DFG), stated in reference to CERN:

> It is the principle of emergence creating science and knowledge which finds its clean expression in such installations, that principle from which about 90% of all great scientific knowledge in the world can be deduced. This principle refers to the innovation as a practically unintended by-product of focussed scientific effort. It remains a hope of humankind that the irreducible higher structure of thinking surfaces with completely new qualities from the structure of which it originated. And it is precisely this which is so difficult to understand, difficult to make it understandable and nevertheless basic for all our actions.[7]

---

[6] Translated by the author from "...es stelle bis jetzt das weitaus sinnvollste dar, was Europa hervorgebracht habe, weil es das scheinbar Sinnloseste sei, im Spekulativen, Abenteuerlichen angesiedelt, in der Neugierde an sich."

[7] Translated from [3]: "Es ist das wissenschafts- und erkenntnisgenerierende Prinzip der Emergenz, das in solchen Anlagen rein zum Ausdruck kommt, jenes Prinzip aus dem rund 90% aller großen wissenschaftlichen Erkenntnis der Welt hergeleitet werden. Dieses Prinzip meint die Innovation als gleichsam unbeabsichtigtes Nebenprodukt gezielter wissenschaftlicher Anstrengung. Daß die höhere Denkstruktur nicht zurückführbar mit völlig neuen Qualitäten aus der Struktur auftaucht, aus der sie entstanden ist, bleibt eine der Hoffnungen der Menschheit. Und eben dies ist es, was so schwer zu verstehen, so schwer verständlich zu machen ist und doch so grundlegend für all unser Tun ist."

# References

1.  Ali-Khalili J (2007) IOP Interactions Dec:8
2.  Schopper H (2003) Phys World Mar:19
3.  Dürrenmatt F (1986) Der Auftrag oder vom Beobachten des Beobachters der Beobachter. Diogenes, Zurich
4.  Frühwald W (1995) Wider die deutsche Angst. Die Zeit 46/1995

# Chapter 12
# CERN – Bringing Nations Together

The acronym 'CERN' has become a trademark for scientific and technical excellence. It is less well known that CERN has made considerable contributions to a better understanding and to building confidence between people from different traditions, mentalities, religions and political systems. The first initiative for the foundation of CERN was started as long ago as 1946 by European physicists who became aware that competition with the USA was only possible by the European countries joining forces [1]. The first discussions started in the frame of UNESCO between Eduardo Amaldi from Italy [2], the French Pierre Auger and Lew Kowarski and the American Isidor Rabi. A second and rather independent initiative was due to the Swiss writer Denis de Rougemont, who had spent the war at Princeton. After his return to Europe in 1948, de Rougemont became, together with Raoul Dautry (Administrateur-General of the French Commissariat l'Energie Atomique, CEA) and other far-sighted diplomats and administrators, one of the driving forces of the 'European Movement' which resulted in the creation of the Centre Européen de la Culture at Lausanne in 1950. The objective was to build bridges between people who had been at war and an international scientific laboratory was considered to be the best 'tool' to bring scientists, administrators and politicians together for peaceful work.

The two initiatives formed the basis for a proposal to the UNESCO General Conference in Florence in June 1950. Rabi, an American Nobel Prize winning physicist, had formulated this enabling motion and later when I invited him to give a speech at the 30th anniversary of CERN he said [3, 4]:

> Europe had been the scene of violent wars... for 200 years. Now we have something new in the founding of CERN, namely Europe has gotten together, in the cause of science.... So I think it is most important for CERN to continue and be the symbol and the driving force of a possible unity of Europe.... I hope that the scientists at CERN will remember that they have other duties than exploring further into particle physics. They represent the combination of centuries of investigation and study,... to show the power of human spirit. So I appeal to them not to consider themselves as technicians... but.... as guardians of this flame of European unity so that Europe can help preserve the peace of the world.

In his speech, Rabi (see Fig. 11.7) also made some remarks which indicated that he considered his involvement in the foundation of CERN as a form of compensation for his contributions to designing the atomic bomb.

H. Schopper, *LEP – The Lord of the Collider Rings at CERN 1980–2000*, 179
DOI 10.1007/978-3-540-89301-1_12, © Springer-Verlag Berlin Heidelberg 2009

Often the argument is made that it is relatively easy to establish cooperation between physicists in basic science since they have the same motivations, communicate using their lingua franca, broken English, and detest secrecy. But activities concerning large facilities as at CERN are not limited to scientists. Administrators have to be involved and when it comes to major decisions even politicians at the highest levels must agree. In this way the spirit of collaboration bred at the level of scientists radiates into many other layers of society. Many examples could be given. During the Cold War CERN was the first organization to conclude an agreement with the Soviet Union, in 1968, establishing a cooperation with the large national laboratory Institute for High Energy Physics (IHEP)[1] at Protvino near Serpukhov. This agreement became a model for a similar agreement between the Soviet Union and the USA and I was told that this in turn facilitated a later agreement between the governments signed by presidents Brezhnev and Ford. Contracts are not worth the paper they written upon unless they are backed by some mutual confidence. Science can help to build up such confidence.

In 1956 the international laboratory Joint Institute for Nuclear Research (JINR) was founded at Dubna northeast of Moscow according to the CERN model for the 'Warsaw pact states' behind the iron curtain. The cooperation between JINR and CERN provided one of the rare bridges for cooperation between physicists from the West and the East during the Cold War. It played a particular role in the cooperation between scientists from West Germany and East Germany. At that time it was the only possibility for scientists of the two parts of Germany to work together and after the unification of Germany it became an important element in knitting new relations between East and West. After the dissolution of of the Soviet Union, JINR received a new task of promoting cooperation among parts of the former Soviet Union. The cooperation between JINR and CERN has even been strengthened during the past few decades, since CERN benefits from the considerable technical competence at JINR and scientists from the East can use the unique CERN facilities. In the meantime, CERN, although formally still a European laboratory, has become a world institution. The LEP experiments were the first step in that direction where all projects and experiments were carried out by multinational groups, including many from non-member states. This development has been brought to a climax by the LHC (see Fig. 12.1).

Indeed some of the collaborations paved the way for international cooperation beyond Europe involving developing countries. A strong cooperation with the China was established. High-energy physicists belonged to the first people who could visit China and meet not only colleagues but also high-level politicians. They were eager to get our advice on their science policy. I visited China for the first time in 1977, immediately after the fall of the 'gang of four' at a time when the roads of Beijing were full of bicycles and practically no cars were visible. In the following years we could observe with astonishment the rapid development of China and its science and it was exciting and for both sides enlightening to come into contact with a

---

[1] At that time IHEP had the best performing proton accelerator, with an energy of 70 GeV.

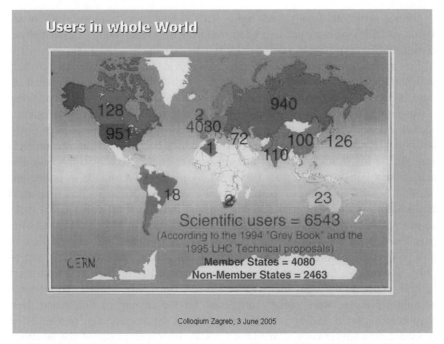

**Fig. 12.1** The worldwide participation of scientists in CERN activities. The *numbers* indicate the users from different countries

completely different culture, both in science and in daily life. One evening I was invited to dinner by the vice-president of an industrial firm which delivered special materials (niobium for superconducting cavities) to CERN. He asked me about the size of CERN and when I posed the same question he answered that his firm was just a small enterprise. When I insisted on knowing the number of employees he remarked, "Just 1.2 million employees." Boasting is not good behaviour in the Far East! This example also shows the different scales which are relevant when comparing Europe with China.

An extraordinary success in bringing people together was achieved by one LEP experiment. The coordinator of the L3 experiment was Samuel Ting, an American Nobel Prize winner of Chinese origin, as mentioned in Chap. 7. His good relations with Deng Xiaoping, the de facto leader of China, made it possible for the first time for Chinese scientists to be allowed to work in a Western country. One day, when I was at DESY, the telephone rang and Ting was at the other end. He asked whether Chinese physicists could come to DESY to participate in PETRA experiments. When I asked where he was, he answered that he was in Deng Xiaoping's office. So I replied, "Yes, they can come, but how many?" A moment of silence to ask Deng and then the following dialogue went on: "About 100." My turn: "That is much too many, what about a dozen?"

Silence again and finally, "All right." That is how they came to DESY in the 1970s. At LEP Ting managed to get approval from the highest authorities that physicists from China and Taiwan could work side by side in the same group.

In some cases scientific collaboration could provide favourable conditions to help individual scientists who had gotten in conflict with their governments. Political authorities were often strongly interested in a good collaboration with CERN. When it was indicating that the cooperation with CERN would be endangered unless a special individual case could be settled in good spirit, it was sometimes possible to achieve a compromise. One example which became well known publicly was Yuri Orlov, a known expert in particle accelerators who as a human-rights activist was jailed in the Soviet Union. Many protests were made in public and the CERN staff also wanted to start a public initiative. Knowing how important face-keeping is for dictatorial regimes, I asked them to wait until I could make an attempt to help Orlov. During a cooperation meeting in Moscow I asked Minister Petrosjansk for a private talk, during which I pointed out that in view of good cooperation we expected that Orlov be allowed to come to CERN. Indeed Orlov could leave the Soviet Union and come to CERN in 1986 and from there he immigrated to the USA and went to Cornell University. Maybe various kinds of public pressure were equally important for the freeing of Orlov and these have been reported widely in the press, whereas the involvement of CERN remained practically unknown. We kept CERN's involvement out of the headlines on purpose since my experience was that, in general, it was easier to help people in difficulties if those in positions of responsibility on the other side could save face.

I learned this on various other occasions. Once a Soviet scientist was elected as a member of the CERN Scientific Policy Committee but he never came to the meetings. When I complained to the director of his institute, whom I knew very well, and warned him that our collaboration might suffer, I was told that the colleague had unfortunately been sick but he was recovering quickly and in future he would attend the meetings. At the next meeting he turned up without any sign of past health problems.

I should like to mention another example which shows how a place such as CERN enjoying the confidence of controversial partners can help. When disarmament negotiations were taking place in Geneva between the USA and the Soviet Union in the wake of the Reagan–Gorbachev summit meeting, at a certain moment the ran into deadlock. One day the head of the US delegation, Alwin Trivelpiece, a physicist whom I knew from earlier collaborations (see Fig. 13.2) called me. He informed me of his worry that the negotiations could come to an abrupt end without a result. He suggested I invite the heads of the two delegations to a dinner at CERN, where in a neutral relaxed atmosphere appreciated by both parties one might hope that a solution could be found. This happened and indeed the negotiations continued with a breakthrough.

Once I invited the ambassadors of the Disarmament Conference in Geneva to visit CERN. At the end of their visit they stated that they had learned that one of the main objectives of CERN was to collide particles, whereas it was their task to avoid collisions between countries. But, they added, CERN is probably doing better, even as far as their objective is concerned.

The most recent offspring of CERN is the SESAME project[2] in the Middle East, a synchrotron radiation laboratory founded under the auspices of UNESCO, following exactly the model of CERN. It is located near Amman in Jordan with present members Bahrain, Cyprus, Egypt, Israel, Iran, Jordan, Palestine, Pakistan and Turkey, and other countries are expected to join. Apart from the promotion of science and technology in the region, the idea was to transfer the spirit of CERN to help create a better climate for peaceful cooperation in a way similar to that which CERN had done for Europe after World War II. When I was asked to be president of the Council of this international intergovernmental organization I simply had to copy the Convention of CERN with only a few small adaptations. The project started with a gift from Germany. A facility for synchrotron radiation at Berlin called BESSY I was closed down after German unification since a larger machine was being built in the eastern part of Berlin, now BESSY II. The components of BESSY I (worth about USD 8 million) were provided free of charge and they became the core of the new project in the Middle East. After extensive discussions with potential users, a final design for a truly competitive facility was decided on. The project has passed the point of no return, the building is finished and the installation of the BESSY I components has started. The personal interest of King Abdullah II of Jordan was essential for the fast realization of the project. The building was inaugurated in November 2008 in his presence and that of Koichiro Matsuura, director-general of UNESCO. The machine is expected to start operating in late 2011. It is very gratifying and for some people incredible to see the representatives of the countries mentioned sitting around the same table discussing very peacefully and objectively the issues concerning the laboratory, in spite of the tense political relations between some of them.

Concluding this chapter, one may say that the efforts made by CERN member states to develop and maintain CERN are not only justified by CERN's enormous contributions to the progress of science and technology but also by helping to create better human and political relations in various parts of society, following the UNESCO slogan 'Science for Peace'.

# References

1. Hermann A, Krige J, Mersits U , Pestre D (1987) History of CERN, vol. 1. North Holland, Amsterdam
2. Schopper H (1995) In: Menzinger F (ed) Italia at CERN, le ragioni di un successo. INFN, Rome
3. (1984) Allocutions prononcées à l'occasion du 30e Anniversaire du CERN, 21 Septembre 1984, brochure CERN. CERN, Geneva
4. Krige J (2004) I.I. Rabi and the birth of CERN. Phys Today 57(9)44

---

[2] See http://www.sesame.org.jo for more information.

# Chapter 13
# The Complicated Transition from LEP to the LHC

The idea that LEP should be followed by a large proton collider, housed in the same tunnel dates back to about 1977. This possibility was studied in detail at a workshop [1] at Lausanne in 1984 and this provided the main arguments for the difficult decision in favour of a LEP tunnel 27 km long (see Chap. 3). This happened in the tradition of CERN to consider and plan new projects at a time when previous ones are not yet finished. This offers the advantage that new projects can be thoroughly discussed with the users' community and that the projects can be designed and prepared in such a way that they can be realized within the time schedules and budgets foreseen.

However, the critical problem concerned the timing of when a previous project should be given a lower priority or even be abandoned in favour of a new project. This became a pertinent question concerning the switch from LEP to a proton collider in the LEP tunnel, the Large Hadron Collider (LHC). Such a decision was further complicated by the US project of a proton–proton collider, the Superconducting Super Collider (SSC), which was in competition with the LHC. Before the SSC was stopped by the US Congress, difficult discussions took place on whether the LHC and the SSC were complementary and both should be built or whether one of them should be cancelled. The fact that the LHC could be realized faster in view of the existing tunnel and other infrastructure at CERN whereas the SSC had to be built 'from scratch' was an important element in the competition.

Hence, it was understandable that Carlo Rubbia was pushing very hard to get a decision in favour of building the LHC at CERN as fast as possible. He had always worked on proton machines, achieved his successes (including the Nobel Prize) using such machines and did not have great affection for electron facilities. His impatience was nurtured by the aforementioned competition with the USA. After the discovery of the $W^+$, $W^-$ and Z particles at CERN in 1983, the USA thought it had lost its leadership in particle physics. The New York Times published the headline "Europe 3, US Not Even Z-Zero" and the US president's science advisor called for the US to "regain leadership" in high-energy physics. As a result the SSC was proposed with a circumference of 87 km and a beam energy of 20 TeV. Acrimonious comparisons between the LHC and the SSC took place in the following years.

H. Schopper, *LEP – The Lord of the Collider Rings at CERN 1980–2000*,
DOI 10.1007/978-3-540-89301-1_13, © Springer-Verlag Berlin Heidelberg 2009

Although some scientific arguments had been given for choosing an energy of 20 TeV, many people were convinced that the real reason for such an energy was that it could not be reached in the LEP tunnel. For proton machines with different energies, a comparison of their discovery potential is not straightforward, since a lower energy can partly be compensated for by higher luminosities (see Chap. 1). No conclusion based on scientific arguments could be achieved as to whether the LHC and the SSC should both be built or not.

At the end of 1985 my first term as director-general of CERN ended and therefore in autumn of 1984 the discussion started on whether my appointment should be extended or a new director-general should be nominated. So far it had not happened that the mandate of a director-general had been prolonged; on the other hand Council delegates wanted to see whether I would keep my promise and finish LEP within the constant budget. Behind the scenes Rubbia was pushing very hard to become the new director-general since he was convinced that only under his leadership would the LHC receive the proper thrust. Because he was one of the few Italian Nobel Prize winners, his candidature received strong support from the Italian government. The result was an embarrassing situation similar to the one at my first nomination. During December 1984 confidential discussions in the Committee of Council took place with the aim of reaching an agreement at its regular meeting on 12 December 1984. Following a proposal by Jean Teillac, 12 delegations, except the Italian delegation, were in favour of extending my mandate until the end of 1988 when LEP would be finished and my retirement age reached. An effort was made to achieve a unanimous decision, but tough negotiations followed which almost led to a diplomatic incident. The Italian ambassador in Bonn complained in a letter to the secretary of state in the German Foreign Ministry (which had no direct responsibility for CERN) about the aggressive attitude of the German delegation in the Committee of Council. The German delegates were J. Rembser, department director from the Federal Ministry for Research and Technology and vice-president of the Council at that time, and Wolfgang Paul from Bonn, a later Nobel Prize winner. Traditionally the German delegation showed a rather restrained attitude when the interests of German citizens were to be defended, but the perception of the Italian government was apparently different. To resolve the impasse, the creation of a working group on the long-term future of CERN was proposed with Rubbia as chairman, with the tacit implication that he would be the main candidate as my successor in 1989. I was asked to elaborate on the details for such a proposal to Council in consultation with Rubbia and various delegates. Finally, at its 77th session on 22 February 1985 the Council decided unanimously to create a Long Range Planning Committee (LRPC) to study the future of CERN after LEP and to extend my mandate until end of 1988.

Rubbia was appointed by Council to chair the working group and its members were nominated by the director-general at the recommendations of Rubbia. In spite of the difficulties in Council, Rubbia and I continued to cooperate amicably and efficiently for the benefit of CERN. The working group had the task of looking at all possible ways to guarantee the future of CERN. A subpanel led by Giorgio Brianti studied with a certain emphasis a proton collider in the LEP tunnel. The LRPC reported to Council [2] in June 1987 and proposed a proton–proton collider, the LHC, with a beam energy of 8 TeV. The idea of a much cheaper proton–antiproton collider was also considered but was abandoned since it is extremely difficult to accumulate sufficient antiprotons to reach useful luminosities. A proton–proton collider needs, on the other hand, two magnetic rings with opposite magnetic fields. For the LHC complicated superconducting magnets with two bores for two beams in opposite directions were developed. Superconducting magnets with a nominal field of 10 T were considered to be necessary. Since such magnets did not exist, a development programme was proposed. The overoptimistic statement was made that "If a decision to construct LHC could be taken in 1989... one would expect first collisions at LHC by 1995"! No definite cost for the total project was quoted. A parallel operation with LEP was envisaged by dividing the year into two operational periods of approximately 5 months for each machine. As proposed already in the Lausanne workshop in 1984, it was foreseen to install the LHC magnets on top of the LEP magnets (Fig. 13.1), which would have also allowed electron–proton collisions, extending the investigations at HERA at DESY to much higher energies.

The next event was the approval of the SSC by President Reagan in January 1987 and in November 1987 a site Waxahachie, in Texas, was chosen. However, the US Congress was reluctant to give the green light to the project, mainly because of the enormous cost. The US Congress urged for the SSC to become an international project with contributions from Europe and Japan, and in order to explore such a possibility we were invited to several hearings of committees of the US Congress. Finally, formal negotiations (Fig. 13.2) were initiated after a G7 head-of-states

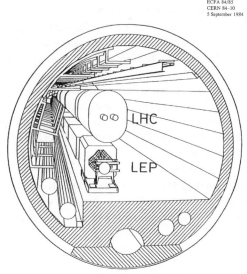

ECFA 84/85
CERN 84–10
5 September 1984

LARGE HADRON COLLIDER
IN THE LEP TUNNEL

**Fig. 13.1** The LHC in the LEP tunnel as proposed in 1984

**Fig. 13.2** Negotiations at the Washington concerning the participation of Europe and Japan in the Superconducting Super Collider (SSC) project. Sitting from *left* to *right*: Josef Rembser (Germany), Alwin Trivelpeace (USA), Harry Atkinson (UK), Tetsuji Nishikawa (Japan). Standing from *left* to *right*: Jule Horowitz (France), Volker Soergel (Germany), Nicola Cabibbo (Italy), Paolo Fasella (EC), Herwig Schopper (CERN). Standing behind from *left* to *right*: Guy Montanet (CERN), Derek Colley (UK), Douglas Stairs (Canada)

meeting at Versailles in France. I participated in 1987 as a member of the European delegation in such talks in Washington where we explored together with Japanese delegates the possibilities of influencing major decisions concerning the parameters of the SSC. The answer we obtained was something like "The President has decided to build such a machine and you have the option to join the project or leave it." This was the end of the SSC as an international project. The USA had still not learned how to organize an international project with equal partners giving them the possibility to participate in decision taking. The efforts to get the SSC approved as a national project failed for technical and financial reasons and because of bad management.[1] On 21 October 1993 the SSC project was permanently cancelled by the US Congress.

This finished the competition between the LHC and the SSC but the interference between the LHC and LEP was not solved. In 1988, my last year as director-general, I presented to Council a medium-term plan for the period 1989–1992 which foresaw an energy upgrade of LEP 1 in three stages to LEP 2:

---

[1] It is not the objective of this book to relate the details of this disaster. See, for example, [3].

1. Manufacture of 32 superconducting cavities in 1990–1991 and their installation in straight section 2.
2. Thirty-two more superconducting cavities to be produced by industry and installed in straight section 6, with possible operation starting in 1992.
3. Depending on the success of the superconducting cavities, 128 cavities could be installed, replacing partly the copper cavities.

In 1989 Rubbia followed me as director-general and he presented to Council in 1990 a long-term strategy paper which foresaw that the LHC would be operating in 1998 in parallel with LEP 2.

In the meantime LEP and its experiments operated well, taking data around the Z particle resonance with 128 copper accelerating cavities providing a total energy gain of 300 MeV per turn. But even when LEP was still on the drawing board, a far-sighted R&D programme for superconducting cavities was initiated to take LEP to higher energies at a later time. The development of superconducting cavities for LEP 2 continued at CERN and in cooperation with other laboratories and industry and became one of the most challenging technology projects ever undertaken at CERN. Superconducting cavities have the advantage that one can obtain high accelerating fields with manageable rf power since heat losses in the cavity walls are greatly reduced (see Sect. 6.4). Two types of cavity were developed. The first type was cavities fabricated out of solid niobium, which is a good superconducting metal at liquid-helium temperatures (about 4 K). The second type consisted of copper cavities which were plated by sputtering on the inside to create a thin film of niobium. This offers the advantage that the amount of expensive niobium is drastically reduced and the heat transport in copper is better than that in in niobium, resulting in better cooling. After extensive development work and experience with the cavities in the LEP beams, preference was given to the copper cavities with niobium films.

In 1990 a new paper "Considerations on the Long-Term Scientific Strategy of CERN" was presented where it was foreseen that 192 copper cavities should be operating in 1994 and that the LHC, still to be installed above LEP, should be commissioned in 1998. Günther Plass and Carlo Wyss played an essential role in this upgrading project.

However, the LEP upgrade got into great difficulties in 1992–1993 because of technical problems with the couplers of the rf cavities and lack of clarity concerning the responsibility for various stages of the production and installation of the cavities. Rubbia played the problems down somewhat since he wanted to avoid negative interference with the approval of the LHC. The situation was remedied when Kurt Hübner got involved in late 1993 and became director of accelerators in 1994. Thanks to the very cooperative attitude of Lyn Evans, project leader of the LHC, is was possible to use part of the cooling system which had been bought for the LHC to cool the LEP rf cavities.

At the beginning of 1994 Rubbia, who I believe would have liked to have seen the approval of the LHC under his direction, was followed as director-general by Christopher Llewellyn-Smith. At the end of 1993 a more realistic long-term plan was presented foreseeing the commissioning of the LHC in 2002 and the realization of LEP 2 in stages starting in late 1995 with 192 copper and 32 superconducting cavities. It was hoped that some budget increases and/or contributions from non-member states could be obtained.

The approval of the LHC turned out to be very difficult, mainly because it was to be realized within a constant budget, a condition well known from LEP. Therefore, further proposals were developed, including a delay in commissioning the LHC until 2003 or 2004 and staging the construction of the detectors. A reduction to a bare minimum of other parts of the CERN programme over the coming years was considered, with a complete closure for 1 year. However, the priority for LEP and its ambitious upgrade was maintained. After the LHC design had been simplified to reduce the cost, the LHC was approved in December 1994, albeit for a two-stage construction with commissioning of stage 1 in 2004 and of stage 2 in 2008. Essential elements for the approval were the existence of the LEP tunnel, the availability of the whole injector system and, last but not least, positive signals from non-member states such as USA, Japan, Russia and India.[2]

At that time the decision was also taken to remove LEP irreversibly from the tunnel since the space above LEP did not seem to be sufficient to install the LHC magnets and a separate cryoline necessary for their cooling (Fig. 13.3). The idea of an electron–proton collider was given up and most of the LEP components were disposed of. However, parts of the infrastructure, such as the cooling system and power equipment, became valuable components of the LHC.

**Fig. 13.3** The LHC in the LEP tunnel

---

[2] The details of the approval of the LHC, e.g. the special difficulties with Germany and the UK, and later budgetary problems will hopefully be described later by another author. A preliminary report was given by Llewellyn-Smith [3].

# References

1. 'Large Hadron Collider in the LEP tunnel', ECFA-CERN Workshop, Lausanne and geneva, 21 to 27 March 1984, ECFA 84/85, CERN 84-102
2. CERN (1987) Report of the Long Range Planning Committee, 27 August 1987. CERN/1658. CERN, Geneva
3. Llewellyn-Smith C (2007) How the LHC came to be. Nature 448:281

# Chapter 14
# The Dramatic Last Period of LEP

In 1994 LEP was still considered to be CERN's current flagship programme. Indeed, strong scientific arguments had emerged in favour of a full upgrade, maximizing LEP's potential. At Fermi National Accelerator Laboratory (FNAL) the top quark had been discovered and the indirect determination of its mass by LEP had been fully confirmed. Further precision measurements at LEP had, in addition, greatly strengthened the physics interest in the energy range to which the upgrade would give access. However, the upgrade required additional resources, above all 32 additional superconducting cavities, which had not been included in the previous long- or medium-term plans and which would cost about CHF 10 million per year over 3 years. Fortunately Christopher Llewellyn-Smith managed to convince Council of the importance of the LEP upgrade and it was approved in 1995.

Some superconducting cavities were installed in 1995 and by 1998 a total of 272 superconducting cavities provided an accelerating voltage of 2,700 MeV per turn, allowing LEP to reach a collision energy of 189 GeV (Table 14.1). By May 1999 the final 16 niobium-plated cavities had been installed, bringing the energy to 192 GeV. But there was better to come. The cavities and their cooling had been designed for a peak accelerating field of 6 MeV/m. The rf accelerating fields are produced by high-frequency electric currents in the superconducting niobium in the cavity walls. For dc currents there are no losses in a superconductor, whereas for high frequencies this is not true. These losses heat up the cavity walls. If the temperature rise is too great, the superconductivity will break down, thus limiting the peak accelerating fields. However, the LEP engineers took a bold risk. With the argument that powerful cooling units would be needed anyway for the LHC magnets, funds were made available to upgrade the cooling power for the LEP cavities. This made it possible to push the accelerating field from 6 to 7 MeV/m, 16% beyond the nominal limit. On 2 August 1999, right on cue for LEP's tenth anniversary, the efforts were rewarded when 100-GeV beams were made to collide. In September the energy was taken up another notch to 101 GeV per beam and LEP ran at that energy for the rest of the year.

These technical achievements set the scene for a climax close to the end of LEP. The four experiments produced excellent results and fitting the precision data obtained provided an indirect estimate for the mass of the so much looked for Higgs

H. Schopper, *LEP – The Lord of the Collider Rings at CERN 1980–2000*,
DOI 10.1007/978-3-540-89301-1_14, © Springer-Verlag Berlin Heidelberg 2009

**Table 14.1** The evolution of LEP's accelerating power, see also Fig. 6.9

| Date | No. of copper cavities | No. of niobium-plated cavities | No of solid niobium cavities | Accelerating voltage per turn | Beam energy (GeV) |
|---|---|---|---|---|---|
| 1990 | 128 | 0 | 0 | 300 | 45 |
| November 1995 | 120 | 56 | 4 | 750 | 70 |
| June 1996 | 120 | 140 | 4 | 1,600 | 80.5 |
| October 1996 | 120 | 160 | 16 | 1,900 | 86 |
| May 1997 | 86 | 224 | 16 | 2,500 | 91.5 |
| May 1998 | 48 | 256 | 16 | 2,750 | 94.5 |
| May 1999 | 48 | 272 | 16 | 2,900 | 96 |
| November 1999 | 48 | 272 | 16 | 3,500 | 101 |
| May 2000 | 56 | 272 | 16 | 3,650 | 104.5 |

particle (see Chap. 8), and a surprisingly low mass of about 90 GeV/$c^2$ was predicted (see Fig. 8.11). Thus, LEP entered the most likely energy range for finding this elusive particle, which is the most important missing element of the standard model. Any little further increment in energy would bring new hopes of its discovery. However, increasing the energy by adding accelerating cavities is not very efficient since the synchrotron radiation losses increase sharply with the beam energy (see Chap. 2) and one has to boost the rf power strongly to compensate for these losses.

Of course, energy alone is not sufficient, there must also be an adequate operating time to accumulate a sufficient number of rare events. Hence, the question arose of whether the running time of LEP could be extended beyond the foreseen end in 1999 into 2000. The additional operating cost of LEP in 2000 seemed small compared with the total investment for LEP, but the estimated CHF 50 million could not be found simply within the CERN budget, which was already being squeezed so hard for construction of the LHC. Llewellyn-Smith managed to find a compromise by taking some funds from a special reserve, allowing a small increase of future debts, getting some extra contributions from some member states and finding some funds within CERN, and in June 1998 Council approved the operation of LEP in 2000. In addition, it had been found possible to schedule the civil engineering work for the LHC in such a way that LEP could continue to operate in 2000 without delaying the start-up of the LHC.

Running at a collision energy of 202 GeV, two of the experiments saw an indication of candidates for the Higgs particle. Excited discussions followed and eight old copper cavities which had been stored somewhere were reactivated and installed, taking the beam energies to 104.5 GeV in May 2000. A hectic search for the Higgs particle started. In early September the ALEPH experiment [1] reported three events which looked like a Higgs particle with a mass of about 114 GeV/$c^2$ decaying into hadron jets. Because of this intriguing observation, operation of LEP was extended until the beginning of November 2000. In October, L3 recorded [2] an event with a decay mode different from that predicted for the Higgs particle but corresponding to a similar mass. DELPHI and OPAL found small excesses of events in this mass range but no clear candidates for the Higgs particle. In summary, the evidence for the discovery of a Higgs particle was not clear.

However, by combinination of the results of the four experiments, a solid lower limit for the mass of the Higgs particle of 114.4 GeV/$c^2$ (95% confidence level, for the experts) could be obtained, which is still today, 8 years after the closure of LEP, the most solid information [3] on the mass of the Higgs particle.

Even after the celebration when LEP was closed down, the search for the Higgs particle continued. A special working group evaluated the combined analysis of the four experiments and reported to the LEP Experiments Committee [4] on 3 November 2000. They concluded that the excess events were in agreement with a Higgs particle with a mass of 115 GeV/$c^2$; however, the significance was not completely convincing. The Committee concluded that the chance of a Higgs particle having such a mass was about 50% and, in summary, the Committee considered there was a sizable likelihood for the discovery of the Higgs particle if LEP continued to operate in 2001. However, the Committee also recognized that an extension of the operation of LEP could have a serious impact on the LHC schedule and, in view of this, no consensus for a recommendation for an extension could be obtained. Luciano Maiani, who had become director-general of CERN, decided that the best policy for CERN was to proceed full-speed ahead with the LHC and to close down LEP. He explained the reasons for his decision in an article in the *CERN Courier* [5]. Despite vociferous pressure to keep LEP going in 2001 all CERN authorities, the Scientific Policy Committee, Committee of Council (17 November) and Council supported the decision by the management and at 8 o'clock in the morning of 2 November 2000 LEP was switched off permanently.

This decision provoked fierce reactions. *Physics World*, the journal of the British Physical Society, published an article entitled "Should LEP Have Been Closed?" The German committee for particle physics immediately organized an opinion poll among the German particle physicists involved in LEP or LHC experiments, with the result that 14 of the 15 groups working on LEP were in favour of an extension and 14 of the 23 groups working at the LHC also agreed to a prolonged operation of LEP provided the LHC would not be delayed by more than 2 years. *Scientific American* [6] wrote, "In a move that surprised and dismayed many physicists, one of the world's leading laboratories has chosen not to continue an experiment that showed every sign of being on the verge of discovering an elusive particle that would have placed the capstone on a century of particle physics." Even the CERN Staff Association [7] contested the decision.

In the meantime, the Tevatron at FNAL had gone through an improvement programme and the experiments were able to determine the mass of the top quark with higher precision. Combining this result with all the precision data obtained by LEP led to a better indirect estimate of the mass of the Higgs particle being obtained, a the surprisingly low value of $(76^{+33}_{-24})$ GeV/$c^2$, which lies much below the directly determined lower mass limit obtained from LEP 2. How this discrepancy can be solved is a tormenting question which will perhaps be answered once the Higgs particle is discovered – if it exists! It would be a great pity if in the end the mass of the Higgs particle were to be found inside the energy range reachable by LEP. The magnets of LEP were able to keep electrons in orbit up to an energy of 125 GeV, but more accelerating power would have been needed. Unfortunately the production contracts for accelerating cavities had been stopped in 1998 and in 2000 it was too late to reactivate industry.

The Tevatron has sufficient energy to produce the Higgs particle provided its mass is as low as indicated by the LEP results. Yet, as explained in Chap. 2, the analysis of proton–antiproton collisions is very complicated and no results have yet been obtained at the Tevatron. Of course, it would be extremely embarrassing if the Higgs particle were detected at the Tevatron with a mass that could have been produced by a further upgraded LEP![1] However, there is no doubt that if the Higgs particle exists at all it should be within the reach of the LHC. Scientific research is a very competitive activity, full of exiting surprises and ahead of time one never knows where the gold will be found. Was it right to close LEP in the face of inconclusive data and give priority to the LHC, which was delayed for other reasons? History will be the ultimate judge.

On 9 and 10 October 2000 a celebration was organized at CERN on the occasion of the closing down of LEP. On the first day of this 'LEP Fest' several ministers gave speeches and on the second day the scientific achievements of LEP were summarized. I had been asked to give a review of the history of LEP and started with

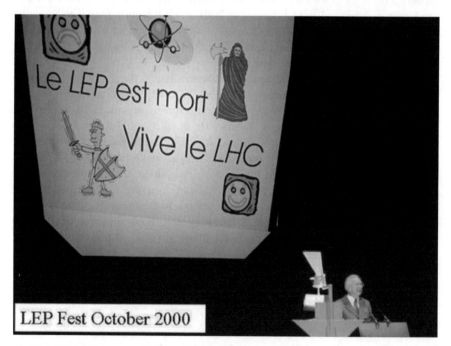

**Fig. 14.1** The LEP Fest at the closure of the facility

---

[1] But this seems to be still some time away [8].

'Le LEP est mort, vive le LHC' paraphrasing the old saying *Le roi est mort, vive le roi* used when a French king died, expressing at the same time sadness about the loss and hope for a new beginning (Fig. 14.1). LEP will be the last circular electron–positron collider.[2] It has provided superb results pushing ahead the frontier of our knowledge of the microcosm and it has also demonstrated that there will be a rich harvest of physics to be reaped by the LHC, whose the cradle it has been. The ministers attending the closing-down celebration unveiled a commemorative plaque which read:

> *We, the participating countries, recognize the outstanding scientific achievements of LEP that have illuminated the family structure of fundamental particles and the texture of our Universe.*
>
> *LEP has stimulated new ideas and technologies with applications reaching far beyond the realms of fundamental physics. Best known is the World Wide Web.*
>
> *LEP has set new standards for international scientific collaboration, giving scientists from all over the world the opportunity to work together and push back the limits of the unknown. Worldwide contacts and relations have been established by using the new instruments and techniques developed at CERN and by the particle physics community.*
>
> *LEP achievements open the way for a new challenge: the Large Hadron Collider (LHC), which will allow us to go deeper in the exploration of the structure of matter, space and time.*

## References

1. Barate R et al (2000) ALEPH collaboration. Observation of an Excess in the Search for the Standard Model Higgs Boson at ALEPH Phys Lett B 495:1
2. Acciari N et al (2000) L3 collaboration. Search for the Standard Model Higgs boson in $e^+e^-$ collisions at s up to 202 GeV Phys Lett B 495:225
3. Grünewald M (2008) Experimental precision tests for the electroweak standard model. In: Hand of particle physics. Springer, Heidelberg
4. LEP Experiments Committee (2000) Minutes of the 56th meeting, 3 November 2000. CERN/LEPC 2000-012; LEPC 56. CERN, Geneva and Search for Standard Model Higgs Boson at LEP at center-of-mass energies of 206 GeV or higher, the four collaborations ALEPH, DELPHI, L3 and OPAL, Nucl. Phys. B, Proc. Supp. 115 (2003) 369–73
5. Maiani L (2001) The road ahead. CERN Cour 41:27
6. Collins GP (2001) Higgs won't fly. Sci Am Feb:12
7. Staff Association du Personnel (2000) CERN staff contest LEP closure: the stakes are of paramount importance for Europe. Bulletin hebdomadaire 47/2000. CERN, Geneva
8. Public Higgs Discovery Group (2008) Tevatron combined high mass Higgs, T. Aaltonen et al., The CDF Collaborations, Phys. Rev. Lett. 100, 211801

---

[2] It may be followed by a linear electron–positron collider which is being discussed as a worldwide project. It will be very expensive and a decision will not be taken before results from the LHC have been obtained.

# Acknowledgments

Many people, scientists, engineers, technicians, administrators and others have contributed to the success of LEP. Without their dedication and efforts LEP could never have been realized. It is impossible to thank all of them individually.

My special thanks go to some colleagues who have helped in editing this book. In particular I should like to thank warmly Emilio Picasso for reading the whole book very carefully. He checked many details, applied many corrections and made a number of suggestions. Sir Chris Llewellyn-Smith gave me many hints concerning the later stages of LEP when he was Director General. Christian Caron from Springer Verlag and the staff were essential in putting the book into its final form.

All the photos were taken from the CERN archives or CERN brochures and CERN has the copyright. Sometimes it was not easy to find them and some were not in a digitized format. Keith Potter and Ray Lewis deserve many thanks for helping in this respect. The quality of some of the figures is therefore not optimal but for a historical report it seemed better to reproduce them as they were used at that time.

I would also like to thank warmly Rolf-Dieter Heuer for contributing the Foreword in spite of his heavy engagement as Director-General of CERN.

Finally my incessant thanks are due to my wife Ingeborg for her steady moral support and her patience.

# Appendix: CERN Organigram 1984

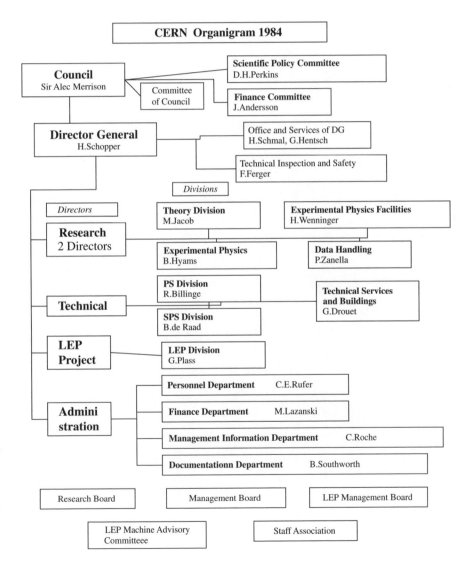

**CERN Organigram 1984**

**Council**
Sir Alec Merrison

**Scientific Policy Committee**
D.H.Perkins

Committee of Council

**Finance Committee**
J.Andersson

**Director General**
H.Schopper

Office and Services of DG
H.Schmal, G.Hentsch

Technical Inspection and Safety
F.Ferger

*Divisions*

*Directors*

**Research**
2 Directors

**Theory Division**
M.Jacob

**Experimental Physics Facilities**
H.Wenninger

**Experimental Physics**
B.Hyams

**Data Handling**
P.Zanella

**Technical**

**PS Division**
R.Billinge

**Technical Services and Buildings**
G.Drouet

**SPS Division**
B.de Raad

**LEP Project**

**LEP Division**
G.Plass

**Administration**

**Personnel Department**     C.E.Rufer

**Finance Department**     M.Lazanski

**Management Information Department**     C.Roche

**Documentationn Department**     B.Southworth

Research Board

Management Board

LEP Management Board

LEP Machine Advisory Committeee

Staff Association

# Appendix: Leading CERN Staff During the LEP Project
(Appointed by CERN Council)

## Directors-General

L. van Hove (Scientific Director-General) and J. Adams (Managing Director-General) (1976–1980)
H. Schopper (1981–1988)
C. Rubbia (1989–1993)
C. Llewellyn-Smith (1994–1998)
L. Maiani (1999–2003)

## Directorate

*Research (two in parallel):*

E. Gabathuler (1981–1986), R. Klapisch (1981–1986), I. Butterworth (1983–1986), J. Thresher (1987–1991), P. Darriulat (1987–1993), W. Hoogland (1989–1993), L. Foà (1994–1998), H. Wenninger (1994–1999), C. Detraz (1999–2003), R. Cashmore (1999–2003)

*Accelerators:* G. Brianti (1981–1989), G. Plass (1990–93), K. Hübner (1994–2003)

*LEP Project Leader:* E. Picasso (1981–1989)

*LHC Project Leader:* L. Evans (since 1994)

*Administration:* H. Heyn (1981–1988), G. Vianès (1989–1991), H. Weber (1992–1994), M. Robbin (1995–2000)

*Human Resources:* G. Martinez (1986–1988)

*Technologies:* H.F. Hoffmann (1990–1999), J. May (1999–2003)

*Informatics:* R. Billinge (1992–1993)

*Forecast and Planning:* C. Roche (1992–1993)

## Division and Department Leaders

*Theory:* J. Prentki (until 1982), M. Jacob (1983–1988), J. Ellis (1989–1991), G. Veneziano (1994–1997), A. de Rujula (1997–2000)

*Experimental Physics*: A. Wetherell (1981–1983), B. Hyams (1984–1987), F. Dydak (1988–1990), J.V. Allaby (1991–1995), G. Goggi (1995–2001)

*Experimental Physics Facilities*: A. Minten (1976–1983), H. Wenninger (1984–1989), P.G. Innocenti (1990–1994), M. Turala (1994–1998)

*Data Handling:* P. Zanella (1981–1988), D.O. Williams (1989–1997), J. May (1997–1999)

*LEP:* G. Plass (1983–1989)

*SPS+LEP:* L. Evans (1990–1993), K.H. Kissler (1994–1999), S. Myers (2000–2003)

*Proton Synchrotron:* G. Munday (1973–1981), R. Billinge (1982–1990), K. Hübner (1991–1993)

*Super Proton Synchrotron:* G. Brianti (1979–1980), B. de Raad (1981–1989), L. Evans (1990–1994), K.H. Kissler (1994–2000)

*Technical Services and Buildings:* H. Laporte (1981–1982), G. Drouet (1983–1985)

*Finance Department:* C. Tièche (1970–1981), M. Lazanski (1982–1988), A. Naudi (1989–2003)

*Personnel Department:* F. Niemann (until 1981), C. Rufer (1982–1986), N. Blackburn (1986–1987), G. Michel (1988–1990), W. Middelkoop (1991–1995), B. Angerth (1996–1998)

*Documentation Department:* B. Southworth (1982–1985)

*Management and Information Department:* H. Roche (1982–1985)

*Technical Inspection and Safety Commission:* F. Ferger (1983–1986), K. Potter (1987–1990), B. de Raad (1991–1996)

This list does not give details on the creation, reorganization or dissolution of divisions and departments. More information can be obtained at http://library.cern.ch/archives/internorg/internalorganization.html.

# Glossary

| | |
|---|---|
| **AdA** | Anello di Accumulazione; the first electron–positron collider built in the 1960s at Frascati National Laboratories, Italy |
| **ACO** | Anneau de Collisions d'Orsay; an electron–positron collider in Orsay, France |
| **ADONE** | The bigger successor of AdA at Frascati National Laboratories |
| **ALEPH** | Apparatus for LEP Physics; one of the four LEP experiments |
| **ALICE** | A Large Ion Collider Experiment; one of the four LHC experiments |
| **ATLAS** | One of the four LHC experiments |
| **BEBC** | Big European Bubble Chamber; built at CERN in the 1970s |
| **CEA** | Commissariat à l'Energie Atomique; a major French organization for research and development, in particular and originally, for atomic and nuclear energy research |
| **CERN** | Conseil Européen pour la Recherche Nucléaire; original name of the European Laboratory for Particle Physics in Geneva |
| **CHEEP** | An electron–positron facility proposed for the SPS at CERN in the 1970s |
| **CMS** | Compact Muon Spectrometer; one of the four experiments at the LHC |
| **DELPHI** | Detector with Lepton, Photon and Hadron Identification; one of the four LEP experiments |
| **DESY** | Deutsches Elektron Synchrotron; research laboratory in Hamburg, Germany |
| **DORIS** | Doppelspeicher Ring-System; an electron–positron collider with two rings at DESY, also used for synchrotron radiation experiments |
| **ECFA** | European Committee for Future Accelerators |
| **ELECTRA** | One of the experiments proposed for LEP but not accepted |
| **EPA** | Electron–Positron Accumulator; part of the beam injection system for LEP |

| | |
|---|---|
| **EUROLEP** | International consortium for the tunnelling of LEP consisting of the companies Impresa Astaldi (Italy), Entrecanales y Tavora (Spain), Fougerolle (France), Philipp Holtzmann (Germany) and Rothpletz Lienhart et Cie (Switzerland) |
| **FKZ** | Forschungszentrum Karlsruhe; at Karlsruhe, Germany |
| **FNAL** | Fermi National Accelerator Laboratory; near Chicago, USA |
| **GLLC** | International consortium for the tunnelling of LEP consisting of the companies C. Baresel (Germany), Chantiers Modernes (France), CSC Impresa Costruzioni (Switzerland), Intrafor-Cofor (France), Locher (Switzerland), and Wayss et Freitag (Germany) |
| **HERA** | Hadron-Elektron-Ring-Anlage; a proton–electron collider at DESY |
| **IHEP** | Institute for High Energy Physics; at Protvino, Russia |
| **ILC** | International Linear Collider; a possible successor to the LHC under discussion as a world project |
| **INFN** | Istituto Nazionale de la Fisica Nucleare; the Italian National Institute supporting nuclear and elementary particle physics |
| **ISR** | Intersecting Storage Rings; the first proton–proton collider at CERN (30 GeV) |
| **JADE** | Japan, Deutschland, England; a compact magnetic detector at PETRA (DESY) |
| **JINR** | Joint Institute for Nuclear Research; in Dubna, Russia, created according to the CERN model for the Warsaw-block states |
| **KEK** | High Energy Accelerator Research Organization, in Japan |
| **L3** | One of the four LEP experiments; named so because of the third submitted letter of intent |
| **LAL** | Laboratoire de l'Accélérateur Linéaire; near Paris, France |
| **LEP** | Large Electron–Positron Collider; at CERN, Switzerland/France |
| **LHC** | Large Hadron Collider; at CERN, Switzerland/France, the largest proton–proton collider ever built, successor to LEP using the LEP tunnel |
| **LHCb** | One of the experiments at the LHC; specialized for B physics |
| **LIL** | LEP Injection Linac; part of the LEP beam injection system |

| | |
|---|---|
| **LOGIC** | A plastic ball detector; one of the experiments proposed for LEP but not accepted |
| **LRPC** | Long Range Planning Committee; created by CERN Council in 1985 |
| **OPAL** | Omni-Purpose Apparatus for LEP; one of the four LEP experiments |
| **PEP** | Positron–Electron Project; a positron–electron collider at SLAC |
| **PETRA** | Positron–Elektron-Tandem-Ring-Anlage; a positron–electron collider with a beam energy of 19 GeV at DESY |
| **PS** | Proton Synchrotron; a proton accelerator at CERN (30 GeV), now also accelerating electrons and ions |
| **QCD** | Quantum chromodynamics. The (quantum field) theory of the strong nuclear force |
| **QED** | Quantum electrodynamics. The (quantum field) theory of the electromagnetic force |
| **RAL** | Rutherford Appleton Laboratory; in the UK |
| **RICH** | Ring imaging Cherenkov counter. A detector giving information on the speed of particles |
| **STAC** | Sampling total absorption counter. A detector component measuring the energy of hadrons, usually called a 'hadron calorimeter' |
| **SC** | Synchro-Cyclotron; the first accelerator at CERN (proton energy of 600 MeV) |
| **SERC** | Science and Engineering Research Council; in the UK |
| **SESAME** | Synchrotron Light for Experimental Science and Applications in the Middle East; a synchrotron radiation laboratory created under the auspices of UNESCO according to the CERN model |
| **SLAC** | Stanford Linear Accelerator Center; in Stanford, California, USA, with a 1-mile-long linear accelerator for electrons |
| **SLC** | Stanford Linear Collider; a positron–electron collider with a beam energy of 50 GeV at SLAC |
| **SPC** | Scientific Policy Committee; advising the CERN Council |

| **SPS** | Super Proton Synchrotron; a proton accelerator at CERN (400 GeV) which started operation in 1976. Later it was transformed into a proton–antiproton collider, leading to the discovery of the W and Z particles in 1983 |
| **SSC** | Superconducting Super Collider; large proton collider with a beam energy of 20 TeV in the USA. Construction was started but stopped by the US Congress |
| **SUSY** | Supersymmetry. An extension of the standard model by requiring a further symmetry between matter particles and particles carrying the forces (interactions) |
| **TEVATRON** | Proton accelerator and storage ring at FNAL, reaching beam energies of 1 TeV |
| **TPC** | Time projection chamber. A detector to reconstruct particle tracks in three dimensions |
| **TRISTAN** | Electron–positron collider at KEK in Japan |
| **UNESCO** | United Nations Educational, Scientific and Cultural Organization |
| **UA1, UA2** | The two experiments at the SPS, discovering the W and Z particles in 1983 |
| **VBA** | Very Big Accelerator; a hypothetical accelerator discussed as a world project |
| **World Wide Web** | A system of extensively linked hypertext documents; developed at CERN |

# Index

Printing: Krips bv, Meppel, The Netherlands
Binding: Stürtz, Würzburg, Germany